I0056366

# METHODS OF HILBERT SPACES IN THE THEORY OF NONLINEAR DYNAMICAL SYSTEMS

# METHODS OF HILBERT SPACES IN THE THEORY OF NONLINEAR DYNAMICAL SYSTEMS

**K. Kowalski**

*Dept. of Biophysics, Institute of Physiology and Biochemistry Lodz, Poland*

**World Scientific**

*Singapore • New Jersey • London • Hong Kong*

*Published by*

World Scientific Publishing Co. Pte. Ltd.
5 Toh Tuck Link, Singapore 596224
*USA office:* 27 Warren Street, Suite 401-402, Hackensack, NJ 07601
*UK office:* 57 Shelton Street, Covent Garden, London WC2H 9HE

**British Library Cataloguing-in-Publication Data**
A catalogue record for this book is available from the British Library.

**METHODS OF HILBERT SPACES IN THE THEORY OF NONLINEAR DYNAMICAL SYSTEMS**

Copyright © 1994 by World Scientific Publishing Co. Pte. Ltd.

*All rights reserved. This book, or parts thereof, may not be reproduced in any form or by any means, electronic or mechanical, including photocopying, recording or any information storage and retrieval system now known or to be invented, without written permission from the publisher.*

For photocopying of material in this volume, please pay a copying fee through the Copyright Clearance Center, Inc., 222 Rosewood Drive, Danvers, MA 01923, USA. In this case permission to photocopy is not required from the publisher.

ISBN-13 978-981-02-1753-2
ISBN-10 981-02-1753-6

*TO MY MOTHER*

# PREFACE

The book is the first monograph on the new powerful method discovered by the author for the study of nonlinear dynamical systems relying on reduction of nonlinear differential equations to the linear, abstract, Schrödinger-like equation in Hilbert space. Besides the possibility of unification of many apparently completely different techniques the introduced "quantal" Hilbert space formalism enables to discover new original methods for solving nonlinear problems arising in investigation of ordinary and partial differential equations as well as difference equations. Applications covered in the book include symmetries and first integrals, linearization transformations, Bäcklund transformations, stroboscopic maps, functional equations involving the case of Feigenbaum-Cvitanovic renormalization equations and chaos.

While it fully draws on the language and technique of standard Hilbert space quantum physics, no background in physics is in principle required for understanding the essential ideas in this book. The necessary explanations for the mathematicians who are not familiar with fundamentals of quantum physics and especially with the Dirac notation which is used throughout this book, are placed in appendices A and B. It is safe to say that the intended audience is composed of physicists, advanced graduate students, and, hopefully, mathematicians who are attracted to the modern theory of nonlinear dynamical systems.

K. KOWALSKI

## CONTENTS

# INTRODUCTION

It is no exaggeration to say that the great area of science and technology owes their development to the theory of differential equations. An example of current interest is a concept of chaos causing a revolution in our understanding of the nature of a deterministic process. Among differential equations the crucial role is played by nonlinear ones. Indeed, whenever the interactions in a physical system or multimolecular reactions in a chemical reactor are taken into consideration then one should deal with nonlinear differential equations.

As our mathematical intuition seems to work better in the case of a linear theory, therefore one would ask about a linear description of nonlinear differential equations. The idea to apply the theory of linear integral equations in the study of nonlinear ordinary differential equations was proposed by Poincaré in 1908. First attempt in this direction was made by Fredholm in 1920. Nevertheless, having in mind the actual state of the theory of linear integral equations he gave up studying of the obtained linear integral equation. In 1931 Koopman [1] reported that one can associate to each Hamiltonian system of classical mechanics a unitary operator satisfying the Schrödinger equation in Hilbert space. The observations of Koopman permit application of the powerful methods of Hilbert spaces in the ergodic theory [2]. The algorithm of embedding of finite nonlinear dynamical systems $d\mathbf{x}/dt = \mathbf{P}(\mathbf{x})$, where $\mathbf{x} \in \mathbf{R}^k$ and $P_i$'s are polynomial in $\mathbf{x}$, into an infinite system of linear differential equations was introduced by Carleman in 1932 [3]. Recently, the Carleman approach which is known today as Carleman linearization or Carleman embedding has been succesfully applied in solving numerous nonlinear problems (see the book [4] and references therein). In their book of 1980 [5] Fučik and Kufner briefly mentioned a method for linearization of nonlinear evolution equations in Banach space relying on replacing some variables in the original vector field by time-dependent parameters. This method was applied in [5] in proving existence theorems for nonlinear ordinary differential equations. In 1982 Wong [6] showed that the Carleman linearization can be viewed as a reduction of nonlinear dynamical systems $d\mathbf{x}/dt = \mathbf{F}(\mathbf{x}, t)$, where $\mathbf{F}$ is analytic in $\mathbf{x}$, to a linear evolution equation in Banach space. Although the significance of the Wong approach for the mathematical foundations of the Carleman embedding might be far-reaching, nevertheless, the Wong formalism is complicated and its use as a practical tool for the study of the concrete nonlinear equations appears limited. In 1988 Wackerbauer [7] demonstrated that the problem of determining the stroboscopic maps for the polynomial finite systems $d\mathbf{x}/dt = \mathbf{P}(\mathbf{x})$ can be brought down to the solution of the infinite linear system of differential equations. The Wackerbauer approach was applied for finding approximate solutions to the systems of nonlinear ordinary differential equations [8]. The method for construction of the Schrödinger equation in Hilbert space which is specified by the nonlinear dynamical system $d\mathbf{x}/dt = \mathbf{F}(\mathbf{x})$, where $\mathbf{F}$ is a differentiable vector field, was described in 1992 by Alanson [9]. Although this method amounts a generalization of the classical Koopman approach, nevertheless the Koopman work [1] was not referenced therein. On the other hand, the obtained Hilbert space formulation for nonlinear dynamical systems was treated by Alanson as a formal property of a particular statistical approach and not as a tool for the study of nonlinear ordinary differential equations. In the above survey of linearization algorithms I did not mentioned treatments which are restricted only to the case of integrable dynamical systems such as, for example, the method of inverse scattering [10]. I also excluded approaches appearing *ad hoc* as a tool for the linearization of a particular nonlinear dynamical system or a narrow class of such systems. Finally, I did not take into consideration the linearization algorithms like Nishigori memory function approach [11] imposing restrictions on the initial data.

The purpose of this book is a systematic, detailed presentation of the Hilbert space approach to the theory of nonlinear dynamical systems discovered in 1985 by the author, relying on reduction of nonlinear ordinary differential equations [12], nonlinear partial differential equations [13], and nonlinear recurrences [14] to the linear, abstract, Schrödinger-like equation in Hilbert space with boson Hamiltonian. The presented Hilbert space formalism amounts a far-reaching generalization of the Carleman embedding. On the one hand, the Carleman linearization corresponds to the particular occupation number representation in the case of ordinary differential equations [12] and difference equations [14]. On the other hand, the actual treatment allows to generalize the Carleman technique to the case of the partial differential equations [13]. Furthermore, the scheme of generalization of the Carleman embedding via the Hilbert space approach can be immediately applied to incorporate the case of the Alanson approach mentioned above. Last but not least, the Hilbert space formalism generalizes the Wackerbauer observations reported earlier. As it has been already mentioned in the preface, the Hilbert space approach described in this book permits application of the whole apparatus of quantum mechanics and quantum field theory in the investigation of classical dynamical systems. This enables unification of many apparently completely different techniques and discovery of new original methods for the study of nonlinear ordinary, partial, and difference equations. Having in mind the applications of the linearization algorithms in the theory of nonlinear dynamical systems, the "quantal" Hilbert space formalism introduced herein seems to be the simplest and the most universal one.

The development of the Hilbert space approach may be described by means of a brief outline of the contents of the four chapters that make up this book. In chapter 1 we introduce the Hilbert space formulation for ordinary differential equations involving discussion of the theory of Lie series, symmetries and first integrals within the actual treatment. We also discuss the alternative methods for linearization which are incorporated by the Hilbert space approach. The same scheme of presentation of the Hilbert space formalism was applied in chapter 2 devoted to the Hilbert space formulation for partial differential equations. The generalization of the treatment to the case of the difference equations is described in chapter 3. The final chapter 4 deals with applications of the Hilbert space approach. All constructions introduced in this book are illustrated by numerous examples. Ends of examples are indicated by the symbol □. Ends of remarks are pointed out by the symbol ■.

The mathematical tools necessary to understand the Hilbert space formalism are summarized in five appendices. As mentioned in the preface, appendices A and B were written for mathematicians to make them familiar with Dirac notation used in this book as well as fundamental notions of quantum physics. Appendix C amounts a survey of the basic properties of boson calculus and coherent states which are crucial for the Hilbert space approach described in the book. The basic facts about position and momentum operators are summarized in appendix D. The final appendix E collects the most important formulae concerning functional derivatives.

Chapter 1

# ORDINARY DIFFERENTIAL EQUATIONS

## 1.1 Evolution equation in Hilbert space

The purpose of this section is to introduce the linear, abstract evolution equation in Hilbert space corresponding to the system of nonlinear ordinary differential equations with analytic nonlinearities. Consider the following nonlinear dynamical system (complex or real):

$$\frac{d\mathbf{z}}{dt} = \mathbf{F}(\mathbf{z}, t), \qquad \mathbf{z}(0) = \mathbf{z}_0, \tag{1.1}$$

where $\mathbf{F}$: $\mathbf{C}^k \times \mathbf{R} \to \mathbf{C}^k$ is analytic in $\mathbf{z}$.

We define the vectors $|z, t\rangle$ in Hilbert space such that

$$|z,t\rangle = e^{\frac{1}{2}(|\mathbf{z}|^2 - |\mathbf{z}_0|^2)}|\mathbf{z}\rangle \tag{1.2a}$$

$$= e^{-\frac{1}{2}|\mathbf{z}_0|^2} e^{\mathbf{z}\cdot\mathbf{a}^\dagger}|\mathbf{0}\rangle, \tag{1.2b}$$

where $|\mathbf{z}\rangle$ is a normalized coherent state (see (C.23)), $\mathbf{z}$ satisfies (1.1), $a_i^\dagger$, $i = 1, \ldots, k$, constituting the vector $\mathbf{a}^\dagger = (a_1^\dagger, \ldots, a_k^\dagger)$ are the standard Bose creation operators (see appendix C), $|\mathbf{0}\rangle$ is the vacuum vector, $\mathbf{u}\cdot\mathbf{v} = \sum_{i=1}^k u_i v_i$, and $|\mathbf{z}|^2 = \mathbf{z}\cdot\mathbf{z}^* = \sum_{i=1}^k |z_i|^2$; $\mathbf{z}^* = (z_1^*, \ldots, z_k^*)$, where the asterisk designates the complex conjugation.

On differentiating both sides of (1.2b) with respect to time and using (C.17) we arrive at the following linear, Schrödinger-like evolution equation in Hilbert space:

$$\frac{d}{dt}|z,t\rangle = M(t)|z,t\rangle, \qquad |z,0\rangle = |\mathbf{z}_0\rangle, \tag{1.3}$$

where the boson "Hamiltonian" is of the form

$$M(t) = \mathbf{a}^\dagger\cdot\mathbf{F}(\mathbf{a}, t). \tag{1.4}$$

Here $\mathbf{a}^\dagger$, $\mathbf{a}$ are the standard Bose creation and annihilation operators, respectively.

Now let $\mathbf{z}(\mathbf{z}_0, t)$ designate the solution of the system (1.1) and $|\mathbf{z}_0, t\rangle$ be the solution of (1.3). Applying $\mathbf{a}$ to both sides of (1.2a) we find that the vectors $|\mathbf{z}_0, t\rangle$ are the eigenvectors of the Bose annihilation operators (coherent states) corresponding to the eigenvalues $\mathbf{z}(\mathbf{z}_0, t)$, i.e.,

$$\mathbf{a}|\mathbf{z}_0, t\rangle = \mathbf{z}(\mathbf{z}_0, t)|\mathbf{z}_0, t\rangle. \tag{1.5}$$

It thus appears that the integration of the nonlinear finite-dimensional system (1.1) is equivalent to the solution of the linear, abstract, Schrödinger-like equation in Hilbert space (1.3).

REMARK 1. One may ask about the time evolution of the normalized coherent states $|\mathbf{z}\rangle$, where $\mathbf{z}$ fulfils (1.1). It follows easily from (C.23) that these vectors satisfy

$$\frac{d}{dt}|\mathbf{z}\rangle = M'(t)|\mathbf{z}\rangle, \qquad |\mathbf{z}(0)\rangle = |\mathbf{z}_0\rangle, \tag{1.6}$$

where

$$M'(t) = \mathbf{a}^\dagger\cdot\mathbf{F}(\mathbf{a},t) - \mathrm{Re}(\mathbf{z}^*\cdot\mathbf{F}(\mathbf{z},t)).$$

Clearly, in the case of the complex system (1.1) the "Hamiltonian" $M'(t)$ depends on the solution to (1.1). In other words, the solution of (1.6) requires the knowledge of the solution of (1.1) and therefore, we then do not deal with any linearization of (1.1). On the other hand, in the case of the real system (1.1) such that

$$\frac{d\mathbf{x}}{dt} = \mathbf{F}(\mathbf{x},t), \qquad \mathbf{x}(0) = \mathbf{x}_0, \tag{1.7}$$

where $\mathbf{F}\colon \mathbf{R}^k \times \mathbf{R} \to \mathbf{R}^k$ is analytic in $\mathbf{x}$, we find

$$\frac{d}{dt}|\mathbf{x}\rangle = M'(t)|\mathbf{x}\rangle, \qquad |\mathbf{x}(0)\rangle = |\mathbf{x}_0\rangle, \tag{1.8}$$

where $|\mathbf{x}\rangle$ is a normalized coherent state, $\mathbf{x}$ fulfils (1.7) and the "Hamiltonian" $M'(t)$ is

$$M'(t) = \mathbf{a}^\dagger\cdot\mathbf{F}(\mathbf{a},t) - \mathbf{a}\cdot\mathbf{F}(\mathbf{a},t) = -i\sqrt{2}\,\hat{\mathbf{p}}\cdot\mathbf{F}(\mathbf{a},t),$$

where $\hat{p}_r = i/\sqrt{2}\,(a_r^\dagger - a_r)$, $r = 1, \ldots, k$, are the momentum operators (see appendix D).

The evolution equation (1.8) seems to be more complicated than (1.3). ■

REMARK 2. We note that in the light of the analogies with quantum mechanics, the eigenvalue equation (1.5) suggests that $\mathbf{a}$ is an "observable" in "Schrödinger picture" (see appendix B) representing $\mathbf{z}$. We shall see in the sequel that the "quantization" scheme $\mathbf{z} \to \mathbf{a}$ is an immanent feature of the actual "quantal" approach. ■

We now return to eq. (1.2a). On projecting (1.2a) on a basis vector $|\mathbf{e}_i\rangle$ of the occupation number representation (see appendix C), where $\mathbf{e}_i = (0,\ldots,0,1_i,0,\ldots,0)$ is the unit vector, and a normalized coherent state $|\mathbf{w}\rangle$, respectively and using (C.25) and (C.28a) we arrive at the following useful formulae relating the solution of (1.3) and (1.1):

$$\langle \mathbf{e}_i|\mathbf{z}_0,t\rangle = z_i(\mathbf{z}_0,t)e^{-\frac{1}{2}|\mathbf{z}_0|^2}, \qquad i = 1,\ldots,k, \tag{1.9}$$

$$\langle \mathbf{w}|\mathbf{z}_0,t\rangle = \exp[-\tfrac{1}{2}(|\mathbf{w}|^2 + |\mathbf{z}_0|^2 - 2\mathbf{w}^*\cdot\mathbf{z}(\mathbf{z}_0,t))]. \tag{1.10}$$

EXAMPLE. Consider the following nonlinear dynamical system:

$$\frac{d\mathbf{z}}{dt} = (\mathbf{v}(t)\cdot\mathbf{z})\mathbf{z}, \qquad \mathbf{z}(0) = \mathbf{z}_0. \tag{1.11}$$

The corresponding evolution equation in Hilbert space (1.3) can be written as

$$\frac{d}{dt}|z,t\rangle = M(t)|z,t\rangle, \qquad |z,0\rangle = |\mathbf{z}_0\rangle, \tag{1.12}$$

where the "Hamiltonian" is

$$M(t) = N\mathbf{v}(t)\cdot\mathbf{a}. \tag{1.13}$$

Here $N = \mathbf{a}^\dagger\cdot\mathbf{a}$ is the total number operator (see appendix C). One finds easily that the "Hamiltonian" (1.13) satisfies the following relation:

$$[M(t), M(t')] = 0 \quad \text{for every} \quad t,\, t' \in \mathbf{R}.$$

Therefore, the solution of (1.12) can be written as

$$\begin{aligned} |\mathbf{z}_0, t\rangle &= \exp\left(\int_0^t M(\tau)\,d\tau\right)|\mathbf{z}_0\rangle \\ &= |\mathbf{z}_0\rangle + \sum_{i=1}^{\infty}\frac{1}{i!}\left(\prod_{j=0}^{i-1}(N+j)\right)\left(\int_0^t \mathbf{z}_0\cdot\mathbf{v}(\tau)\,d\tau\right)^i. \end{aligned} \tag{1.14}$$

On using (1.9), (C.8) and (C.25) we arrive at the solution of the system (1.11) such that

$$\begin{aligned} z_i(\mathbf{z}_0, t) &= \langle\mathbf{e}_i|\mathbf{z}_0, t\rangle\exp(\tfrac{1}{2}|\mathbf{z}_0|^2) \\ &= z_{0i}\sum_{j=0}^{\infty}\left(\int_0^t \mathbf{z}_0\cdot\mathbf{v}(\tau)\,d\tau\right)^j \end{aligned} \tag{1.15}$$

The solution (1.15) can be analytically continued to

$$z_i(\mathbf{z}_0, t) = \frac{z_{0i}}{1 - \int_0^t \mathbf{z}_0\cdot\mathbf{v}(\tau)\,d\tau}, \qquad i = 1, \ldots, k. \quad \square \tag{1.16}$$

We now show that the Carleman embedding technique is included by the introduced Hilbert space formalism as a special case. We begin by recalling the Carleman embedding technique [4]. Consider the nonlinear dynamical system (complex or real) of the form

$$\frac{d\mathbf{z}}{dt} = \mathbf{F}(\mathbf{z}, t), \qquad \mathbf{z}(0) = \mathbf{z}_0, \tag{1.17}$$

where $\mathbf{F}\colon \mathbf{C}^k \times \mathbf{R} \to \mathbf{C}^k$ is analytic in $\mathbf{z}$. On setting

$$z_{\mathbf{n}} = \prod_{i=1}^{k} (z_i(t))^{n_i},$$

where $\mathbf{z}(t)$ satisfies (1.17) and $\mathbf{n} \in \mathbf{Z}_+^k$ (here $\mathbf{Z}_+$ is the set of nonnegative integers), we arrive at the following linear differential-difference equation:

$$\frac{dz_{\mathbf{n}}}{dt} = \sum_{\mathbf{n}' \in \mathbf{Z}_+^k} \tilde{M}_{\mathbf{nn}'}(t) z_{\mathbf{n}'}. \tag{1.18}$$

Evidently, the solution of the system (1.17) is linked to the solution of (1.18) by

$$z_i = z_{\mathbf{e}_i}, \qquad i = 1, \dots, k,$$

where $\mathbf{e}_i = (0, \dots, 0, 1_i, 0, \dots, 0)$ is the unit vector.

This means that the finite-dimensional nonlinear system (1.17) is embedded into the infinite-dimensional linear system resulting from (1.18) when one introduces an order in $z_{\mathbf{n}}$, $\tilde{M}_{\mathbf{nn}'}$.

We now return to eq. (1.3). This equation written in the occupation number representation (see appendix C) takes the form of the differential-difference equation

$$\frac{dz_{\mathbf{n}}}{dt} = \sum_{\mathbf{n}' \in \mathbf{Z}_+^k} M_{\mathbf{nn}'}(t) z_{\mathbf{n}'}, \tag{1.19a}$$

$$z_{\mathbf{n}}(0) = \left( \prod_{i=1}^{k} \frac{z_{0i}^{n_i}}{\sqrt{n_i!}} \right) \exp(-\tfrac{1}{2}|\mathbf{z}_0|^2), \tag{1.19b}$$

where $z_{\mathbf{n}}(t) = \langle \mathbf{n} | z, t \rangle$ and $M_{\mathbf{nn}'}(t) = \langle \mathbf{n} | M(t) | \mathbf{n}' \rangle$.

The following relation is an immediate consequence of (1.2a) and (C.25):

$$z_{\mathbf{n}}(t) = \left( \prod_{i=1}^{k} \frac{(z_i(t))^{n_i}}{\sqrt{n_i!}} \right) \exp(-\tfrac{1}{2}|\mathbf{z}_0|^2). \tag{1.20}$$

By (1.20) (see also (1.9)) the solution of the system (1.1) and the solution of (1.19) are related by

$$z_i(t) = z_{\mathbf{e}_i}(t) \exp(\tfrac{1}{2}|\mathbf{z}_0|^2), \qquad i = 1, \dots, k. \tag{1.21}$$

Thus, it turns out that the Carleman embedding technique corresponds to the particular occupation number representation in the actual, canonical, basis independent Hilbert space formalism. The ansatz (1.20) coincides up to the multiplicative constant with the ansatz introduced by Steeb [15] who recognized for the first time that the Carleman embedding matrix can be written with the help of the Bose operators.

REMARK. We note that the Carleman linearization within the introduced "quantal" approach plays the same role as the Heisenberg matrix mechanics in quantum mechanics (see appendix B). In particular, it is interesting to find out that both the Carleman embedding and Heisenberg matrix mechanics were the earliest versions of the corresponding theories. ∎

The advantage of the present approach is that it enables one to translate the methods and the notions of quantum mechanics into the language of the theory of nonlinear dynamical systems. Motivated by the importance of the concept of path integrals in solving numerous problems in quantum physics [16] we now investigate the meaning of this notion within the introduced "quantum-mechanical" Hilbert space formalism.

Consider the following autonomous nonlinear dynamical system:

$$\frac{d\mathbf{z}}{dt} = \mathbf{F}(\mathbf{z}), \qquad \mathbf{z}(0) = \mathbf{z}_0, \tag{1.22}$$

where $\mathbf{F}$: $\mathbf{C}^k \to \mathbf{C}^k$ is analytic in $\mathbf{z}$.

The corresponding evolution equation (1.3) in Hilbert space takes the form

$$\frac{d}{dt}|z,t\rangle = M|z,t\rangle, \qquad |z,0\rangle = |\mathbf{z}_0\rangle, \tag{1.23}$$

where $M = \mathbf{a}^\dagger \cdot \mathbf{F}(\mathbf{a})$.

The formal solution of (1.23) is

$$|\mathbf{z}_0, t\rangle = V(t)|\mathbf{z}_0\rangle,$$

where the evolution operator $V(t)$ is given by

$$V(t) = e^{tM}.$$

In other words, the integration of the system (1.22) is equivalent to calculating the evolution operator $V(t)$. The kernel of the evolution operator $V(t)$ in the coherent-state representation can be written as

$$\langle\mathbf{w}|V(t)|\mathbf{z}_0\rangle = \langle\mathbf{w}|e^{tM}|\mathbf{z}_0\rangle = \langle\mathbf{w}|\mathbf{z}_0, t\rangle,$$

where $|\mathbf{w}\rangle$ is a normalized coherent state.

Taking into account (C.17) and (C.28a) we find that for infinitesimal $\Delta t$,

$$\begin{aligned}\langle\mathbf{w}|e^{\Delta tM}|\mathbf{z}_0\rangle &= \langle\mathbf{w}|(1+\Delta tM)|\mathbf{z}_0\rangle \\ &= \exp[-\tfrac{1}{2}(|\mathbf{w}|^2 + |\mathbf{z}_0|^2 - 2\mathbf{w}^*\cdot\mathbf{z}_0) + \Delta t\mathbf{w}^*\cdot\mathbf{F}(\mathbf{z}_0)].\end{aligned} \tag{1.24}$$

Now we proceed as usual in the case of a path integral. For a finite time $t$ we devide it in $n$ subintervals at length $\Delta t$, where $\Delta t$ is small enough to apply (1.24). Using (1.24) and (C.29) it follows that

$$I_n = \langle\mathbf{w}|\mathbf{z}_0, n\Delta t\rangle = \langle\mathbf{w}|e^{n\Delta tM}|\mathbf{z}_0\rangle \tag{1.25a}$$

$$= \int \left(\prod_{i=1}^{n-1} d\mu(\mathbf{w}_i)\right) \langle \mathbf{w}|e^{\Delta t M}|\mathbf{w}_{n-1}\rangle\langle \mathbf{w}_{n-1}|e^{\Delta t M}|\mathbf{w}_{n-2}\rangle \cdots \langle \mathbf{w}_2|e^{\Delta t M}|\mathbf{w}_1\rangle\langle \mathbf{w}_1|e^{\Delta t M}|\mathbf{z}_0\rangle \tag{1.25b}$$

$$= \int \left(\prod_{i=1}^{n-1} d\mu(\mathbf{w}_i)\right)$$
$$\times \exp\left(\frac{1}{2}(\mathbf{w}_n^*\cdot\mathbf{w}_n - \mathbf{w}_0^*\cdot\mathbf{w}_0) + \sum_{r=1}^{n}[-\mathbf{w}_r^*\cdot(\mathbf{w}_r - \mathbf{w}_{r-1}) + \Delta t \mathbf{w}_r^*\cdot\mathbf{F}(\mathbf{w}_{r-1})]\right), \tag{1.25c}$$

where we set $\mathbf{w}_n = \mathbf{w}^*$ and $\mathbf{w}_0 = \mathbf{z}_0$.

The formal limit $\Delta t \to 0$, $n \to \infty$ of (1.25c) is the path integral such that

$$\begin{aligned} I &= \lim_{n\to\infty} I_n \\ &= \int D\mu(\mathbf{u}) \exp[\tfrac{1}{2}(\mathbf{u}^*(t)\cdot\mathbf{u}(t) - \mathbf{u}^*(0)\cdot\mathbf{u}(0))] \exp\left[\int_0^t d\tau\, \mathbf{u}^*\cdot\left(-\frac{d\mathbf{u}}{d\tau} + \mathbf{F}(\mathbf{u})\right)\right], \end{aligned}$$

where $\mathbf{u}^*(t) = \mathbf{w}^*$ and $\mathbf{u}(0) = \mathbf{z}_0$. It must be borne in mind that in the above integral the variables $\mathbf{u}^*(t)$, $\mathbf{u}(t)$ and $\mathbf{u}^*(0)$, $\mathbf{u}(0)$ are independent ones. Indeed, $\mathbf{u}(t)$ and $\mathbf{u}^*(0)$ are the integration variables whereas $\mathbf{u}^*(t)$ and $\mathbf{u}(0)$ are fixed.

Evidently, the integrals $I_n$ can be treated as successive approximations for the path integral $I$. These approximations correspond via (1.25a) to successive approximations of the solution to (1.23) given by the recursive relation

$$\begin{aligned} |\mathbf{z}_0, (n+1)\Delta t\rangle &= e^{\Delta t M}|\mathbf{z}_0, n\Delta t\rangle \\ &= (1 + \Delta t M)|\mathbf{z}_0, n\Delta t\rangle, \qquad n \in \mathbf{Z}_+. \end{aligned} \tag{1.26}$$

Furthermore, projecting (1.26) onto the vector $|\mathbf{e}_i\rangle$ and using (1.9) we find that the recurrence (1.26) corresponds to the following scheme of successive approximations for the solution of the dynamical system (1.22):

$$\mathbf{z}_{n+1} = \mathbf{z}_n + \Delta t \mathbf{F}(\mathbf{z}_n), \qquad n \in \mathbf{Z}_+, \tag{1.27}$$

where $\mathbf{z}_n = \mathbf{z}(\mathbf{z}_0, n\Delta t)$.

On calculating $I_n$ given by (1.25c) with the help of (C.38) or simply projecting (1.26) onto the normalized coherent state $|\mathbf{w}\rangle$ and using (1.10) we arrive at the following formula relating successive approximations $I_n$ for the path integral $I$ and successive approximations of the solution $\mathbf{z}(\mathbf{z}_0, t)$ to the system (1.22):

$$I_n = \exp[-\tfrac{1}{2}(|\mathbf{w}|^2 + |\mathbf{z}_0|^2 - 2\mathbf{w}^*\cdot\mathbf{z}_n)],$$
$$I = \langle \mathbf{w}|\mathbf{z}_0, t\rangle = \lim_{n\to\infty} I_n = \exp[-\tfrac{1}{2}(|\mathbf{w}|^2 + |\mathbf{z}_0|^2 - 2\mathbf{w}^*\cdot\mathbf{z}(\mathbf{z}_0, t))].$$

We have thus shown that the path integral within the "quantal" Hilbert space formulation corresponds to the classical Euler procedure of successive approximations given by (1.27) (Euler's broken-line method).

## 1.2 Operator evolution equations

As we have already demonstrated above, the advantage of the actual treatment is that it allows to apply the whole apparatus of quantum mechanics in the study of nonlinear dynamical systems. We now illustrate this observation by an example of the "Heisenberg picture" and "interaction picture" within the introduced "quantal" approach. Consider the Schrödinger-like equation (1.3). This equation is equivalent to the following operator evolution equation:

$$\frac{dV}{dt} = M(t)V, \qquad V(0) = I, \tag{1.28}$$

where $V(t)$ is the evolution operator satisfying

$$|\mathbf{z}_0, t\rangle = V(t)|\mathbf{z}_0\rangle. \tag{1.29}$$

Here $|\mathbf{z}_0, t\rangle$ is the solution of (1.3). It thus appears that the the integration of the nonlinear dynamical system (1.1) is equivalent to the solution of the linear operator evolution equation in Hilbert space (1.28). The formal solution to (1.28) can be written in the form (see formulae (B.11) and (B.12)):

$$V(t) = \exp[\Phi(t)], \tag{1.30}$$

where the "phase operator" is given by the Magnus expansion

$$\Phi(t) = \int_0^t d\tau\, M(\tau) + \frac{1}{2}\int_0^t d\tau_2 \int_0^{\tau_2} d\tau_1\, [M(\tau_2), M(\tau_1)] + \cdots. \tag{1.31}$$

Consider now the "Heisenberg equations of motion" which satisfy the time-dependent Bose annihilation operators

$$\frac{d\mathbf{a}(t)}{dt} = [\mathbf{a}(t), M_{\rm H}(t)], \qquad \mathbf{a}(0) = \mathbf{a}, \tag{1.32}$$

where $M_{\rm H}(t)$ is the "Hamiltonian" in "Heisenberg picture" (see appendix B):

$$M_{\rm H}(t) = V(t)^{-1}M(t)V(t),$$

where $V(t)$ is the evolution operator given by (1.28).

The formal solution to (1.32) is

$$\mathbf{a}(t) = V(t)^{-1}\mathbf{a}V(t). \tag{1.33}$$

Hence, taking into account the relation

$$\mathbf{a}(t)|\mathbf{z}_0\rangle = \mathbf{z}(\mathbf{z}_0, t)|\mathbf{z}_0\rangle, \tag{1.34}$$

which is an immediate consequence of (1.5), (1.29) and (1.33), and making use of (1.30) and (B.16) we get the formal solution to the system (1.1)

$$\mathbf{z}(\mathbf{z}_0,t) = \langle \mathbf{z}_0|\mathbf{a}(t)|\mathbf{z}_0\rangle \tag{1.35a}$$

$$= \mathbf{z}_0 + \sum_{i=1}^{\infty}\frac{(-1)^i}{i!}\langle \mathbf{z}_0|[\Phi(t),\ldots,[\Phi(t),\mathbf{a}]\ldots]|\mathbf{z}_0\rangle. \tag{1.35b}$$

REMARK. We note that in view of (1.35a) the solutions of the dynamical systems (1.1) are the covariant symbols (see appendix C) of the time-dependent Bose annihilation operators. More precisely, the following relation holds:

$$\mathbf{a}(t) = \mathbf{z}(\mathbf{a},t), \tag{1.36}$$

where $\mathbf{z}(\mathbf{z}_0,t)$ is the solution to (1.1). It might also be observed that whenever the "quantization" scheme $\mathbf{z}\to\mathbf{a}$ mentioned above is assumed, where $\mathbf{z}$ satisfies (1.1), then (1.35a) forms the "Ehrenfest's theorem" (see appendix B) within the presented approach. ■

EXAMPLE. Consider the system (1.11). The corresponding "phase operator" is given by

$$\Phi(t) = \int_0^t d\tau\, M(\tau) = N\int_0^t d\tau\, \mathbf{v}(\tau)\cdot\mathbf{a}.$$

It is easy to verify that

$$\underbrace{[\Phi(t),\ldots,[}_{n-\text{times}}\Phi(t),\mathbf{a}]\ldots] = (-1)^n n!\left(\int_0^t d\tau\,\mathbf{v}(\tau)\cdot\mathbf{a}\right)^n \mathbf{a}. \tag{1.37}$$

The solution (1.15) to the system (1.11) follows immediately from (1.37) and (1.35b). □

Consider now the case of the autonomous system (1.1)

$$\frac{d\mathbf{z}}{dt} = \mathbf{F}(\mathbf{z}), \qquad \mathbf{z}(0) = \mathbf{z}_0, \tag{1.38}$$

where $\mathbf{F}\colon \mathbf{C}^k \to \mathbf{C}^k$ is analytic in $\mathbf{z}$.

The "phase operator" reduces then to

$$\Phi(t) = tM,$$

where $M = \mathbf{a}^{\dagger}\cdot\mathbf{F}(\mathbf{a})$, and the solution of (1.38) given by (1.35b) takes the form of the formal power series in $t$

$$\mathbf{z}(\mathbf{z}_0,t) = \mathbf{z}_0 + \sum_{i=1}^{\infty}\frac{(-t)^i}{i!}\langle \mathbf{z}_0|[M,\ldots,[M,\mathbf{a}]\ldots]|\mathbf{z}_0\rangle. \tag{1.39}$$

By applying (C.2a) one can easily check that (1.39) coincides with the formal solution of (1.38) given by Lie series [17]. We note that in the case of the nonautonomous analytic system (1.1) with $\mathbf{F}$ analytic in $\mathbf{x}$ and $t$ one also obtains the expansion of the form (1.39). Indeed, every $k$-dimensional nonautonomous system (1.1) can be converted to the $(k+1)$-dimensional autonomous one by identifying $t$ from the right-hand side of (1.1) with an auxiliary variable $x_{k+1}$ (we pass to the extended phase space [18]). Such autonomous system is of the form

$$\begin{aligned} \frac{d\mathbf{x}}{dt} &= \mathbf{F}(\mathbf{x}, x_{k+1}), \\ \frac{dx_{k+1}}{dt} &= 1, \qquad \mathbf{x}(0) = \mathbf{x}_0, \; x_{k+1}(0) = 0. \end{aligned}$$

The corresponding "Hamiltonian" is given by

$$M = \mathbf{a}^{\dagger}\cdot\mathbf{F}(\mathbf{a}, a_{k+1}) + a_{k+1}^{\dagger}.$$

Thus, it turns out that the technique of Lie series corresponds to the particular "Heisenberg picture" within the introduced "quantal" Hilbert space approach to the theory of nonlinear dynamical systems. On the other hand, it appears that (1.35b) amounts a generalization of the technique of Lie series to the case when the vector field from the right-hand side of (1.1) is not analytic in $t$.

We now demonstrate that the classical formulation of the Lie series approach corresponds to the particular Bargmann representation (see appendix C) in the introduced canonical Hilbert space formalism. Consider the real dynamical system

$$\frac{d\mathbf{x}}{dt} = \mathbf{F}(\mathbf{x}), \qquad \mathbf{x}(0) = \mathbf{x}_0, \tag{1.40}$$

where $\mathbf{F}\colon \mathbf{R}^k \to \mathbf{R}^k$ is analytic in $\mathbf{x}$.

Taking into account (1.9) we find that the solution to (1.40) can be expressed by

$$\begin{aligned} x_i(\mathbf{x}_0, t) &= \langle \mathbf{e}_i | e^{tM} | \mathbf{x}_0 \rangle e^{\frac{1}{2}\mathbf{x}_0^2} \\ &= \sum_{r=0}^{\infty} \frac{t^r}{r!} \langle \mathbf{e}_i | M^r | \mathbf{x}_0 \rangle e^{\frac{1}{2}\mathbf{x}_0^2}, \qquad i = 1, \ldots, k. \end{aligned} \tag{1.41}$$

We now introduce the vectors $|i, r\rangle$ such that

$$|i, r\rangle = (M^{\dagger})^r |\mathbf{e}_i\rangle, \qquad r \in \mathbf{Z}_+, \tag{1.42}$$

where $M^{\dagger}$ is the Hermitian conjugate of $M$:

$$M^{\dagger} = \mathbf{F}(\mathbf{a}^{\dagger})\cdot\mathbf{a}.$$

Evidently, the vectors (1.42) obey the following recurrence in Hilbert space:

$$|i, r+1\rangle = M^{\dagger}|i, r\rangle, \qquad |i, 0\rangle = |\mathbf{e}_i\rangle. \tag{1.43}$$

On writing (1.43) in the Bargmann representation (see formulae (C.31) and (C.34)) we arrive at the following relation:

$$\tilde{\phi}_{ir+1}(\mathbf{x}_0) = \mathbf{F}(\mathbf{x}_0)\cdot\frac{\partial}{\partial \mathbf{x}_0}\tilde{\phi}_{ir}(\mathbf{x}_0), \qquad \tilde{\phi}_{i0}(\mathbf{x}_0) = x_{0i}, \qquad i = 1, \ldots, k, \tag{1.44}$$

where

$$\tilde{\phi}_{ir}(\mathbf{x}_0) = \langle \mathbf{x}_0|i,r\rangle e^{\frac{1}{2}\mathbf{x}_0^2} = \langle \mathbf{e}_i|M^r|\mathbf{x}_0\rangle e^{\frac{1}{2}\mathbf{x}_0^2}. \tag{1.45}$$

An obvious solution of (1.44) is

$$\tilde{\phi}_{ir}(\mathbf{x}_0) = \left(\mathbf{F}(\mathbf{x}_0)\cdot\frac{\partial}{\partial \mathbf{x}_0}\right)^r x_{0i}, \qquad i = 1, \ldots, k.$$

Hence, using (1.45) and (1.41) we finally obtain

$$\mathbf{x}(\mathbf{x}_0, t) = \sum_{r=0}^{\infty}\frac{t^r}{r!}\left(\mathbf{F}(\mathbf{x}_0)\cdot\frac{\partial}{\partial \mathbf{x}_0}\right)^r \mathbf{x}_0. \tag{1.46}$$

We have thus shown that the standard formulation for the Lie series approach based upon (1.46) corresponds to the particular Bargmann representation within the canonical Hilbert space formalism. On the other hand, it turns out that the problem of the solution of (1.40) can be reduced to the solution of the linear, abstract recurrence in Hilbert space (1.43). We note that (1.43) holds true also in the case of the complex autonomous systems.

We now demonstrate by an exampe of the Riccati system of the projective type that the counterpart of the "interaction picture" within the introduced "quantal" Hilbert space approach (see apendix B) is the classical method of variation of constants.

EXAMPLE. Consider the real system

$$\frac{d\mathbf{x}}{dt} = L\mathbf{x} + (\mathbf{c}(t)\cdot\mathbf{x})\mathbf{x}, \qquad \mathbf{x}(0) = \mathbf{x}_0, \tag{1.47}$$

where $L$: $\mathbf{R}^k \to \mathbf{R}^k$ is a linear operator. The corresponding evolution equation (1.3) can be written as

$$\frac{d}{dt}|x,t\rangle = (M_0 + M_1(t))|x,t\rangle, \qquad |x,0\rangle = |\mathbf{x}_0\rangle, \tag{1.48}$$

where $M_0 = \mathbf{a}^\dagger\cdot L\mathbf{a}$ and $M_1(t) = N\mathbf{c}(t)\cdot\mathbf{a}$; $N = \mathbf{a}^\dagger\cdot\mathbf{a}$ is the total number operator (see appendix C). Passing to the "interaction picture"

$$|\widetilde{x,t}\rangle = \exp(-tM_0)|x,t\rangle, \tag{1.49}$$

we obtain the following evolution equation:

$$\frac{d}{dt}|\widetilde{x,t}\rangle = \tilde{M}_1(t)|\widetilde{x,t}\rangle, \qquad |\widetilde{x,0}\rangle = |\mathbf{x}_0\rangle, \tag{1.50}$$

where $\tilde{M}_1(t)$ is the "Hamiltonian" in the "interaction picture" such that

$$\tilde{M}_1(t) = \exp(-tM_0)M_1(t)\exp(tM_0).$$

Now, using (B.16) and

$$[M_0, M_{1\mathbf{c}}] = -M_{1\tilde{L}\mathbf{c}}, \tag{1.51}$$

where $M_{1\mathbf{v}} = N\mathbf{v}\cdot\mathbf{a}$ and $\tilde{L}$ designates the transpose of $L$, we get

$$\tilde{M}_1(t) = N\tilde{\mathbf{v}}(\mathbf{c}(t), t)\cdot\mathbf{a}, \tag{1.52}$$

where $\tilde{\mathbf{v}}(\mathbf{x}_0, t)$ is the solution of the linear part of the system (1.47) with the operator $L$ replaced by its transpose. The nonlinear dynamical system corresponding to (1.50), where $\tilde{M}_1(t)$ is given by (1.52), is of the form

$$\frac{d\tilde{\mathbf{x}}}{dt} = (\tilde{\mathbf{v}}(\mathbf{c}(t), t)\cdot\tilde{\mathbf{x}})\tilde{\mathbf{x}}, \qquad \tilde{\mathbf{x}}(0) = \mathbf{x}_0. \tag{1.53}$$

We note that the solution of (1.53) has been already obtained (see formulae (1.11) and (1.16)). Hence, taking into account (1.2a) we find that the solution $\widetilde{|\mathbf{x}_0, t\rangle}$ to (1.50) can be written as

$$\widetilde{|\mathbf{x}_0, t\rangle} = e^{\frac{1}{2}(\tilde{\mathbf{x}}(t)^2 - \mathbf{x}_0^2)}|\tilde{\mathbf{x}}(t)\rangle = \varphi(t)|\tilde{\mathbf{x}}(t)\rangle,$$

where $\tilde{\mathbf{x}}(t) = \mathbf{x}_0/(1 - \int_0^t \mathbf{x}_0\cdot\tilde{\mathbf{v}}(\mathbf{c}(\tau), \tau)\, d\tau) = \mathbf{x}_0/(1 - \int_0^t \mathbf{v}(\mathbf{x}_0, \tau)\cdot\mathbf{c}(\tau)\, d\tau)$ is the solution to (1.53); $\mathbf{v}(\mathbf{x}_0, t)$ is the solution to the linear part of (1.47). Furthermore, eq. (1.49) yields

$$|x, t\rangle = \varphi(t)e^{tM_0}|\tilde{\mathbf{x}}(t)\rangle. \tag{1.54}$$

A look at (1.54) is enough to conclude that the solution $\mathbf{x}(\mathbf{x}_0, t)$ to (1.47) is given by (see (1.5)):

$$\begin{aligned} \mathbf{x}(\mathbf{x}_0, t) &= e^{tL}\tilde{\mathbf{x}}(t) && (1.55\text{a}) \\ &= \frac{\mathbf{v}(\mathbf{x}_0, t)}{1 - \int_0^t \mathbf{c}(\tau)\cdot\mathbf{v}(\mathbf{x}_0, \tau)\, d\tau} && (1.55\text{b}) \end{aligned}$$

As an immediate consequence of (1.55a), we find that the above procedure of integration of the system (1.47) is equivalent to the method of variation of constants. □

The reader would have noticed that the commutation relation (1.51) describing an authomorphism of the Lie algebra generated by $M_0$ and $M_{1\mathbf{c}}$ has played the crucial role. In fact, it has enabled one to calculate the "Hamiltonian" $\tilde{M}_1(t)$ in a closed form. With this remark serving as our motivation we now describe the evolution operator technique of solving the nonlinear dynamical systems (1.1) such that the corresponding "Hamiltonian" (1.4) is linear in generators of a Lie algebra. This technique follows directly from the method for finding the evolution operator for the

Schrödinger equation introduced in ref. [19] and the actual treatment which allows to translate the methods familiar from quantum mechanics into the language of the theory of nonlinear dynamical systems. We illustrate the algorithm by an example of the system (1.47).

EXAMPLE. Consider the system (1.47). The "Hamiltonian" for (1.48) can be written as

$$M(t) = M_0 + \mathbf{c}(t)\cdot\mathbf{M},$$

where $\mathbf{M} = N\mathbf{a}$. It is easy to verify that the operators $M_0$, $M_i$, $i = 1, \ldots, k$, form the Lie algebra. We have

$$[M_0, M_i] = -\sum_{j=1}^{k} L_{ij} M_j, \qquad [M_i, M_j] = 0. \tag{1.56}$$

Now we recall that the solution of the system (1.47) is equivalent to the solution of the operator evolution equation

$$\frac{dV}{dt} = M(t)V, \qquad V(0) = I. \tag{1.57}$$

Following [19] we seek $V(t)$ in the form

$$V(t) = \exp[\alpha(t)M_0]\exp[\boldsymbol{\beta}(t)\cdot\mathbf{M}], \tag{1.58}$$

where $\alpha(t)$, $\beta_i(t)$, $i = 1, \ldots, k$, are functions of time such that $\alpha(0) = 0$, $\beta_i(0) = 0$. On differentiating both sides of (1.58) with respect to time and using the operator identity which follows directly from (B.16) and (1.56)

$$e^{\alpha M_0}\mathbf{M}e^{-\alpha M_0} = e^{-\alpha L}\mathbf{M},$$

we arrive at the following relation:

$$\frac{dV}{dt} = \left(\frac{d\alpha}{dt}M_0 + \frac{d\boldsymbol{\beta}}{dt}\cdot e^{-\alpha L}\mathbf{M}\right)V. \tag{1.59}$$

Comparing the two sides of (1.57) and (1.59) we obtain

$$\begin{aligned}\frac{d\alpha}{dt} &= 1,\\ \frac{d\boldsymbol{\beta}}{dt} &= e^{\alpha \tilde{L}}\mathbf{c}(t), \qquad \alpha(0) = 0,\ \boldsymbol{\beta}(0) = \mathbf{0}.\end{aligned} \tag{1.60}$$

The solution of the trivial system (1.60) is

$$\alpha(t) = t, \qquad \boldsymbol{\beta}(t) = \int_0^t e^{\tau\tilde{L}}\mathbf{c}(\tau)\, d\tau,$$

where $\tilde{L}$ is the transpose of $L$. Now, using (1.58), (1.14) and reasoning as in the final part of the previous example we obtain the solution (1.55b).

REMARK. We note that the linearity of the "Hamiltonian" in the generators of a Lie algebra is not sufficient for integrability of a nonlinear dynamical system. A counterexample is the one-dimensional Riccati equation

$$\frac{dx}{dt} = \alpha(t)x^2 + \beta(t)x + \gamma. \tag{1.61}$$

Indeed, the "Hamiltonian" $M(t)$ corresponding to (1.61) is

$$M(t) = \alpha(t)Na + \beta(t)N + \gamma a^\dagger, \tag{1.62}$$

where $N = a^\dagger a$ is the number operator. Further, we have

$$[Na, N] = Na, \qquad [Na, a^\dagger] = 2N, \qquad [N, a^\dagger] = a^\dagger, \tag{1.63}$$

that is the "Hamiltonian" (1.62) is linear in the generators of the Lie algebra (1.63). Nevertheless, (1.61) is well-known to be integrable only in exceptional cases. Applying the above algorithm in the case of the nonintegrable Riccati equation we arrive at the other nonintegrable one satisfied by one of the coefficients in the ansatz of the form (1.58). ■

## 1.3 Symmetries and first integrals

This section discusses the symmetries and first integrals for nonlinear dynamical systems within the Hilbert space approach [20]. We first study the symmetries. Consider the real autonomous system

$$\frac{d\mathbf{x}}{dt} = \mathbf{F}(\mathbf{x}), \tag{1.64}$$

where $\mathbf{F}\colon \mathbf{R}^k \to \mathbf{R}^k$ is analytic in $\mathbf{x}$.

We recall that the vector field $\boldsymbol{\sigma}(\mathbf{x})$ is a symmetry of (1.64) if it leaves (1.64) invariant up to the order $\epsilon$, that is the equation

$$\frac{d\mathbf{x}'}{dt} = \mathbf{F}(\mathbf{x}'), \tag{1.65}$$

where $\mathbf{x}' = \mathbf{x} + \epsilon\boldsymbol{\sigma}$, must be correct within order $\epsilon$. Using (1.65) and

$$\frac{d\boldsymbol{\sigma}}{dt} = \mathbf{F}(\mathbf{x})\cdot\frac{\partial}{\partial\mathbf{x}}\boldsymbol{\sigma}, \qquad \mathbf{F}(\mathbf{x} + \epsilon\boldsymbol{\sigma}) = \mathbf{F}(\mathbf{x}) + \epsilon\boldsymbol{\sigma}\cdot\frac{\partial}{\partial\mathbf{x}}\mathbf{F} + \mathcal{O}(\epsilon^2),$$

it follows that the symmetry $\boldsymbol{\sigma}$ satisfies

$$\boldsymbol{\sigma}\cdot\frac{\partial}{\partial\mathbf{x}}\mathbf{F} = \mathbf{F}\cdot\frac{\partial}{\partial\mathbf{x}}\boldsymbol{\sigma}. \tag{1.66}$$

On introducing the Lie bracket of analytic vector fields $\mathbf{f}$, $\mathbf{g}$: $\mathbf{R}^k \to \mathbf{R}^k$ such that

$$[\mathbf{f}, \mathbf{g}] = \mathbf{g}\cdot\frac{\partial}{\partial \mathbf{x}}\mathbf{f} - \mathbf{f}\cdot\frac{\partial}{\partial \mathbf{x}}\mathbf{g} \tag{1.67}$$

and assuming that $\boldsymbol{\sigma}$ is analytic in $\mathbf{x}$, we find that the condition (1.66) can be written in the form

$$[\mathbf{F}, \boldsymbol{\sigma}] = 0.$$

We now return to the Hilbert space formalism. Consider the mapping

$$\mathbf{f} \to L_{\mathbf{f}} = \mathbf{a}^\dagger\cdot\mathbf{f}(\mathbf{a}), \tag{1.68}$$

where $\mathbf{f}$: $\mathbf{R}^k \to \mathbf{R}^k$ is analytic. Eqs. (1.67) and (C.2a) taken together yield

$$[L_{\mathbf{f}}, L_{\mathbf{g}}] = L_{[\mathbf{f},\mathbf{g}]}. \tag{1.69}$$

This means that (1.68) establishes isomorphism between Lie algebra of analytic vector fields with the bracket (1.67) and Lie algebra of boson operators linear in creation operators. Using formulae (1.68) and (1.69) we can write the condition for $\boldsymbol{\sigma}$ to be a symmetry of the system (1.64) in the form

$$[M, \Sigma] = 0, \tag{1.70}$$

where $M = \mathbf{a}^\dagger\cdot\mathbf{F}(\mathbf{a})$ is the "Hamiltonian" corresponding to the system (1.64) and

$$\Sigma = \mathbf{a}^\dagger\cdot\boldsymbol{\sigma}(\mathbf{a}). \tag{1.71}$$

We have thus shown that the symmetries of the nonlinear dynamical system (1.64) correspond within the "quantal" Hilbert space approach to operators of the form (1.71) commuting with the "Hamiltonian".

REMARK. Notice that in the particular case of linear vector fields the formulae (1.68) and (1.69) take the form

$$A \to L_A = \mathbf{a}^\dagger\cdot L\mathbf{a}, \tag{1.72a}$$

$$[L_A, L_B] = L_{[A,B]}, \tag{1.72b}$$

where $A$, $B$: $\mathbf{R}^k \to \mathbf{R}^k$ are the linear operators and $[A, B] = AB - BA$. The map (1.72a) is usually known as the *Jordan map* [21]. ■

We now demonstrate that the standard theory of symmetries for classical dynamical systems (1.64) based on the concept of a vector field is included by the canonical Hilbert space formalism as the particular case of the Bargmann representation. Consider equation (1.66). Using the commutator of first-order differential operators

$$X_{\mathbf{f}} = \mathbf{f}(\mathbf{x})\cdot\frac{\partial}{\partial \mathbf{x}}, \qquad X_{\mathbf{g}} = \mathbf{g}\cdot\frac{\partial}{\partial \mathbf{x}}, \tag{1.73}$$

where $\mathbf{f}$, $\mathbf{g}$: $\mathbf{R}^k \to \mathbf{R}^k$ are analytic in $\mathbf{x}$ (they are also called analytic vector fields), such that

$$[X_{\mathbf{f}}, X_{\mathbf{g}}] = \left(\mathbf{f}\cdot\frac{\partial}{\partial \mathbf{x}}\mathbf{g} - \mathbf{g}\cdot\frac{\partial}{\partial \mathbf{x}}\mathbf{f}\right)\cdot\frac{\partial}{\partial \mathbf{x}}, \tag{1.74}$$

we find that the condition (1.66) can be written in the following equivalent form:

$$[X_{\mathbf{F}}, X_{\boldsymbol{\sigma}}] = 0, \tag{1.75}$$

where $X_{\mathbf{F}} = \mathbf{F}(\mathbf{x})\cdot\frac{\partial}{\partial \mathbf{x}}$ and $X_{\boldsymbol{\sigma}} = \boldsymbol{\sigma}(\mathbf{x})\cdot\frac{\partial}{\partial \mathbf{x}}$. Note that $X_{\boldsymbol{\sigma}}$ is the infinitesimal generator of the transformation $\mathbf{x}' = \mathbf{x} + \epsilon\boldsymbol{\sigma}$. The operator $X_{\boldsymbol{\sigma}}$ is also called the symmetry of (1.64). Let us introduce the following mapping:

$$X_{\mathbf{f}} \to L_{\mathbf{f}}^{\dagger} = \mathbf{f}(\mathbf{a}^{\dagger})\cdot\mathbf{a}. \tag{1.76}$$

Taking into account (C.2b) it follows that under the mapping (1.76) the commutator (1.74) transforms as

$$[X_{\mathbf{f}}, X_{\mathbf{g}}] \to [L_{\mathbf{f}}^{\dagger}, L_{\mathbf{g}}^{\dagger}]. \tag{1.77}$$

This shows that (1.76) establishes the isomorphism of the Lie algebra of first-order differential operators with analytic coefficients (analytic vector fields) and the Lie algebra of boson operators linear in annihilation operators. As an immediate consequence of (1.76) and (1.77), we find that the symmetry condition (1.75) can be simply written as the conjugate of (1.70), that is,

$$[M^{\dagger}, \Sigma^{\dagger}] = 0, \tag{1.78}$$

where $M^{\dagger} = \mathbf{F}(\mathbf{a}^{\dagger})\cdot\mathbf{a}$ and $\Sigma^{\dagger} = \boldsymbol{\sigma}(\mathbf{a}^{\dagger})\cdot\mathbf{a}$. Evidently, (1.78) coincides with (1.75) in the case of the Bargmann representation (see appendix C) when the Bose operators take the form

$$\mathbf{a}^{\dagger} = \mathbf{x}, \qquad \mathbf{a} = \frac{\partial}{\partial \mathbf{x}}. \tag{1.79}$$

In other words, the standard theory of symmetries for dynamical systems based on the concept of a vector field corresponds to the particular Bargmann representation within the introduced canonical Hilbert space approach.

REMARK. An easy inspection shows that the mapping

$$\mathbf{f} \to X_{\mathbf{f}}$$

defines the anti-isomorphism of the Lie algebra of vector fields with the bracket (1.67) and the Lie algebra of vector fields with the bracket (1.74), i.e.,

$$[X_{\mathbf{f}}, X_{\mathbf{g}}] = X_{-[\mathbf{f},\mathbf{g}]}.$$

Analogously, the mapping

$$\mathbf{f} \to L_{\mathbf{f}}^{\dagger} = \mathbf{f}(\mathbf{a}^{\dagger})\cdot\mathbf{a}$$

is the anti-isomorphism of the Lie algebra of vector fields with the bracket (1.67) and the Lie algebra of boson operators linear in annihilation operators, that is,

$$[L_{\mathbf{f}}^{\dagger}, L_{\mathbf{g}}^{\dagger}] = L^{\dagger}_{-[\mathbf{f},\mathbf{g}]}. \quad \blacksquare$$

We now examine the first integrals for nonlinear dynamical systems within the Hilbert space formalism. Consider the real autonomous system

$$\frac{d\mathbf{x}}{dt} = \mathbf{F}(\mathbf{x}), \tag{1.80}$$

where $\mathbf{F}\colon \mathbf{R}^k \to \mathbf{R}^k$ is analytic in $\mathbf{x}$.

Recall that the time-dependent first integrals $I(\mathbf{x},t)$ for (1.80) obey

$$\frac{\partial}{\partial t}I(\mathbf{x},t) + \mathbf{F}(\mathbf{x})\cdot\frac{\partial}{\partial \mathbf{x}}I(\mathbf{x},t) = 0. \tag{1.81}$$

This means that the total time derivative of $I(\mathbf{x},t)$ along the solution of (1.80) vanishes, i.e.,

$$\frac{d}{dt}I(\mathbf{x},t) = 0.$$

It follows immediately from (1.81) that whenever the first integral does not depend explicitly on time, then (1.81) reduces to

$$\mathbf{F}(\mathbf{x})\cdot\frac{\partial}{\partial \mathbf{x}}I(\mathbf{x}) = 0, \tag{1.82}$$

that is the derivative of $I(\mathbf{x})$ in the direction of the vector field $\mathbf{F}$ (the Lie derivative) equals zero. Let us now specialize to the first integrals that are analytic in spatial variables. On using (C.2a) we arrive at the following equivalent form of (1.81) and (1.82), respectively:

$$\partial_t I(\mathbf{a},t) + [I(\mathbf{a},t), M] = 0, \tag{1.83}$$

$$[M, I(\mathbf{a})] = 0, \tag{1.84}$$

where $M = \mathbf{a}^{\dagger}\cdot\mathbf{F}(\mathbf{a})$ is the "Hamiltonian" corresponding to the system (1.80).

Thus, it turns out that the first integrals within the Hilbert space approach are represented by operator invariants which are functions of Bose annihilation operators only. As an illustration of this observation, we now derive with the help of the actual treatment $k$ functionally independent local (time-dependent) first integrals $I_r(\mathbf{x},t)$, $r = 1, \ldots, k$, for the $k$-dimensional system (1.80). Consider the Bose operators such that

$$\mathbf{I}(\mathbf{a},t) = e^{tM}\mathbf{a}e^{-tM}. \tag{1.85}$$

The operators $I_r(\mathbf{a},t)$, $r = 1, \ldots, k$, are easily seen to satisfy (1.83). Thus, the functions $I_r(\mathbf{x},t)$, $r = 1, \ldots, k$, are the first integrals of the system (1.80). Moreover, these integrals are functionally independent since the Jacobian det $\frac{\partial \mathbf{I}}{\partial \mathbf{x}}$ does not vanish identically (the matrix $\frac{\partial \mathbf{I}}{\partial \mathbf{x}} = [\frac{\partial I_i}{\partial x_j}]$ is the unit matrix at $t = 0$). It thus appears that the problem of determining the first integrals reduces to the calculation of (1.85). Proceeding as in the section 1.2 (see formula (1.39)) we find that $k$ time-dependent first integrals $I_r(\mathbf{x}, t)$, $r = 1, \ldots, k$, of the system (1.80) are given by the following formal power series in $t$:

$$\mathbf{I}(\mathbf{x}, t) = \langle \mathbf{x}|\mathbf{I}(\mathbf{a}, t)|\mathbf{x}\rangle = \mathbf{x} + \sum_{i=1}^{\infty} \frac{t^i}{i!}\langle \mathbf{x}|[M, \ldots, [M, \mathbf{a}] \ldots]|\mathbf{x}\rangle,$$

where $|\mathbf{x}\rangle$ is a normalized coherent state.

As with the ordinary Lie series, the above expression coincides with the generalized Lie series expansion [22] for the $k$ functionally independent first integrals of the system (1.80).

REMARK. The operators $\mathbf{I}(\mathbf{a}, t)$ are integrals of motion, therefore they are isospectrally deformed during the time evolution, i.e. the eigenvectors of operators $\mathbf{I}(\mathbf{a}, t)$ do not depend on time. In fact, equations (1.85) and (C.17) taken together yield

$$\mathbf{I}(\mathbf{a}, t)|\mathbf{z}_0, t\rangle = \mathbf{z}_0|\mathbf{z}_0, t\rangle, \tag{1.86}$$

where $|\mathbf{z}_0, t\rangle = e^{tM}|\mathbf{z}_0\rangle$ is the solution of the evolution equation in Hilbert space

$$\frac{d}{dt}|z, t\rangle = M|z, t\rangle, \qquad |z, 0\rangle = |\mathbf{z}_0\rangle, \tag{1.87}$$

corresponding to the nonlinear dynamical system (complex or real)

$$\frac{d\mathbf{z}}{dt} = \mathbf{F}(\mathbf{z}), \qquad \mathbf{z}(0) = \mathbf{z}_0. \tag{1.88}$$

It is noteworthy that the equation (1.86) can be interpreted as the "quantal" infinite-dimensional counterpart of the Lax equation. In fact, the integrals $I_r$, $r = 1, \ldots, k$, are obviously in involution

$$[I_r, I_s] = 0, \qquad r, s = 1, \ldots, k. \tag{1.89}$$

These integrals are also functionally independent. Indeed, the relation (1.85) taken at $t = 0$ and the fact that the Bose annihilation operators $a_r$, $r = 1, \ldots, k$, are independent ones, imply

$$f(I_1, \ldots, I_k)|_{t=0} = f(a_1, \ldots, a_k) \neq 0, \tag{1.90}$$

where $f$ is an arbitrary nonvanishing identically analytic function. In our opinion, the "complete integrability" of (1.87) manifested by (1.89) and (1.90) means that one can brought down the dynamics of the evolution equation in Hilbert space (1.87) to the dynamics of the classical finite-dimensional dynamical system (1.88) ■

We now derive the vector equations in Hilbert space corresponding to (1.83) and (1.84). Consider eq. (1.83). Let us define the vectors $|\phi, t\rangle$ such that

$$|\phi, t\rangle = I(\mathbf{a}^\dagger, t)|\mathbf{0}\rangle. \tag{1.91}$$

On taking the Hermitian conjugate of (1.83) and using (1.91) we find that the vectors (1.91) satisfy the following evolution equation in Hilbert space:

$$\frac{d}{dt}|\phi, t\rangle = -M^\dagger|\phi, t\rangle. \tag{1.92}$$

The time-dependent first integrals of (1.80) and the solutions to (1.92) are related by (see (C.31) and (C.32)):

$$I(\mathbf{x}, t) = \langle \mathbf{x}|\phi, t\rangle e^{\frac{1}{2}\mathbf{x}^2},$$

where $|\mathbf{x}\rangle$ is a normalized coherent state.

Consider now eq. (1.84). Proceeding analogously as in the case of eq. (1.83) we introduce the vectors

$$|\phi\rangle = I(\mathbf{a}^\dagger)|\mathbf{0}\rangle.$$

These vectors obey the following equation in Hilbert space:

$$M^\dagger|\phi\rangle = 0. \tag{1.93}$$

The first integrals of (1.80) that do not depend explicitly on time are linked to solutions of (1.93) by

$$I(\mathbf{x}) = \langle \mathbf{x}|\phi\rangle e^{\frac{1}{2}\mathbf{x}^2}, \tag{1.94}$$

where $|\mathbf{x}\rangle$ is a normalized coherent state.

We have thus shown that the problem of finding analytic time-dependent first integrals can be reduced to the solution of an abstract evolution equation in Hilbert space (1.92). This equation is applied in section 4.1.1, where the method for determining first integrals of (1.80) with exponential time dependence is introduced. On the other hand, it turns out that the problem of determining analytic first integrals that do not depend explicitly on time can be brought down to finding the kernel for a linear boson operator in Hilbert space, that is solving eq. (1.93). It should also be noted that the standard approach to the theory of first integrals for dynamical systems utilizing (1.81) and (1.82) corresponds to the particular Bargmann representation within the introduced canonical Hilbert space formalism. In fact, proceeding as in the case of symmetries (see (1.79)) we find that (1.92) and (1.93) written in the Bargmann representation coincide with (1.81) and (1.82), respectively.

REMARK. As a consequence of (1.91), (1.92) and (1.85), we see that the problem of determining $k$ functionally independent first integrals of (1.80) reduces to solving the following system of abstract evolution equations in Hilbert space:

$$\frac{d}{dt}|r,t\rangle = -M^{\dagger}|r,t\rangle, \qquad |r,0\rangle = |\mathbf{e}_r\rangle, \tag{1.95}$$

where $|r,t\rangle = I_r(\mathbf{a}^{\dagger},t)|\mathbf{0}\rangle$ and $\mathbf{e}_r = (0,\ldots,0,1_r,0,\ldots,0)$ are the unit vectors.

The $k$ functionally independent first integrals of (1.80) and the solution to (1.95) are related by

$$I_r(\mathbf{x},t) = \langle \mathbf{x}|r,t\rangle e^{\frac{1}{2}\mathbf{x}^2}, \qquad r = 1,\ldots,k. \quad \blacksquare$$

EXAMPLE. Consider the following dynamical system:

$$\begin{aligned} \frac{dx_1}{dt} &= x_3, \\ \frac{dx_2}{dt} &= x_4, \\ \frac{dx_3}{dt} &= -\frac{1-\mu}{3}x_1^3 - x_1x_2^2, \\ \frac{dx_4}{dt} &= -\frac{1-\mu}{3}x_2^3 - x_2x_1^2. \end{aligned} \tag{1.96}$$

The system (1.96) is the Hamiltonian one. In fact, it is easy to verify that the Hamiltonian is given by

$$H(\mathbf{x}) = \frac{1}{2}(x_3^2 + x_4^2) + \frac{1}{2}x_1^2x_2^2 + \frac{1-\mu}{12}(x_1^4 + x_2^4).$$

We now derive the second first integral for the system (1.96). Consider eq. (1.96). The Hermitian conjugate of the operator $M$ corresponding to (1.96) is of the form

$$M^{\dagger} = a_3^{\dagger}a_1 + a_4^{\dagger}a_2 - \left(\frac{1-\mu}{3}a_1^{\dagger 3} + a_1^{\dagger}a_2^{\dagger 2}\right)a_3 - \left(\frac{1-\mu}{3}a_2^{\dagger 3} + a_2^{\dagger}a_1^{\dagger 2}\right)a_4.$$

By specializing to the analytic first integrals and assuming that we can separate in the first integral the coordinate and momentum variables we make the following ansatz:

$$|\phi\rangle = \sum_{n_1n_2} \alpha_{n_1n_2}|n_1n_200\rangle + \sum_{n_3n_4} \beta_{n_3n_4}|00n_3n_4\rangle, \tag{1.97}$$

where the vectors $|\mathbf{n}\rangle$, $\mathbf{n} \in \mathbf{Z}_+^4$, span the occupation number representation (see appendix C). Inserting (1.97) into (1.93) yields

$$\mu = 0,$$

$$|\phi\rangle = \alpha_{13}(|1300\rangle + |3100\rangle) + \sqrt{\frac{3}{2}}\alpha_{13}|0011\rangle,$$

where $\alpha_{13} = \alpha_{31}$, $\beta_{11} = \sqrt{\frac{3}{2}}\alpha_{13}$. Hence, putting $\alpha_{13} = \sqrt{6}$ and using (1.94) together with (C.25) we finally obtain the following first integral of (1.96):

$$I(\mathbf{x}) = x_1x_2^3 + x_1^3x_2 + 3x_3x_4. \quad \square$$

## 1.4 Alternative linearization approaches

### *1.4.1 Wackerbauer linearization*

As the title of this section indicates, we shall discuss the alternative methods for linearization of nonlinear dynamical systems generalized by the actual treatment. All of them were briefly mentioned in the introduction. We examine first the Wackerbauer approach [7,8]. Consider the real autonomous system

$$\frac{d\mathbf{x}}{dt} = \mathbf{F}(\mathbf{x}), \qquad \mathbf{x}(0) = \mathbf{x}_0, \tag{1.98}$$

where $\mathbf{F}$: $\mathbf{R}^k \to \mathbf{R}^k$ is analytic in $\mathbf{x}$.

We recall that if $\mathbf{x}(t)$ satisfies (1.98), then for fixed $T \in \mathbf{R}$, a mapping $\mathbf{S}(\cdot, T)$: $\mathbf{R}^k \to \mathbf{R}^k$ defined by

$$\mathbf{S}(\mathbf{x}(t), T) = \mathbf{x}(t+T) \tag{1.99}$$

is called the *stroboscopic map* of *strobe time* $T$. The stroboscopic map is usually applied in approximate solution of (1.98). Indeed, using the group property of solutions to (1.98)

$$\mathbf{x}(\mathbf{x}_0, t_1 + t_2) = \mathbf{x}(\mathbf{x}(\mathbf{x}_0, t_1), t_2),$$

where $\mathbf{x}(\mathbf{x}_0, t)$ is the solution of (1.98), we find that the $n$th iterate of $\mathbf{S}(\cdot, T)$ ("stroboscopic picture" of the trajectory) satisfies

$$\mathbf{S}_T^n(\mathbf{x}_0) = \mathbf{x}(\mathbf{x}_0, nT),$$

where $\mathbf{S}_T(\mathbf{x}_0) \equiv \mathbf{S}(\mathbf{x}_0, T)$. Therefore, if one finds an approximate stroboscopic map for small strobe time $T$, then for a finite time $t$ we devide it in $n$ small subintervals at length $T$ and obtain by iterating $\mathbf{S}_T$ an approximate solution of (1.98). Anyway, the problem of finding stroboscopic map is as difficult as solving the original system (1.98). In fact, by virtue of (1.98) and (1.99) we have

$$\frac{d}{dT}\mathbf{S}(\mathbf{x}_0, T) = \mathbf{F}(\mathbf{S}(\mathbf{x}_0, T)), \qquad \mathbf{S}(\mathbf{x}_0, 0) = \mathbf{x}_0. \tag{1.100}$$

The crucial point of the Wackerbauer approach was to make the following ansatz for the stroboscopic map:

$$S_i(\mathbf{x}_0, T) = \sum_{\mathbf{n}\in \mathbf{Z}_+^k} c_{i\mathbf{n}}(T) \prod_{j=1}^{k} x_{0j}^{n_j}, \qquad i = 1, \ldots, k. \tag{1.101}$$

Assuming that $\mathbf{F}$ is polynomial in $\mathbf{x}$ and substituting (1.101) into (1.100) she found that the coefficients $c_{i\mathbf{n}}(T)$ in the expansion (1.101) obey the infinite system of nonlinear ordinary differential equations which due to the hierarchical dependence of the modes $c_{i\mathbf{n}}$ reduces to the infinite-dimensional linear one if $\mathbf{F}(\mathbf{0}) = \mathbf{0}$.

We now demonstrate that the Hilbert space formalism introduced in this book includes the Wackerbauer linearization as a special case. Consider the real nonlinear dynamical system

$$\frac{d\mathbf{x}}{dt} = \mathbf{F}(\mathbf{x}, t), \qquad \mathbf{x}(0) = \mathbf{x}_0, \tag{1.102}$$

where $\mathbf{F}$: $\mathbf{R}^k \times \mathbf{R} \to \mathbf{R}^k$ is analytic in $\mathbf{x}$.

Recall that the corresponding Schrödinger-like equation in Hilbert space is of the form

$$\frac{d}{dt}|x, t\rangle = M(t)|x, t\rangle, \qquad |x, 0\rangle = |\mathbf{x}_0\rangle, \tag{1.103}$$

where $M(t) = \mathbf{a}^\dagger \cdot \mathbf{F}(\mathbf{a}, t)$ and $|\mathbf{x}_0\rangle$ is a normalized coherent state, and that the formal solution of (1.103) is given by

$$|\mathbf{x}_0, t\rangle = V(t)|\mathbf{x}_0\rangle, \tag{1.104}$$

where $V(t)$ is the evolution operator defined by

$$\frac{dV}{dt} = M(t)V, \qquad V(0) = I. \tag{1.105}$$

We also recall that the solution $\mathbf{x}(\mathbf{x}_0, t)$ to (1.102) and the solution $|\mathbf{x}_0, t\rangle$ of (1.103) are related by

$$x_i(\mathbf{x}_0, t) = \langle \mathbf{e}_i | \mathbf{x}_0, t\rangle e^{\frac{1}{2}\mathbf{x}_0^2}, \qquad i = 1, \ldots, k, \tag{1.106}$$

where $|\mathbf{e}_i\rangle$ is a basis vector of the occupation number representation (see appendix C) and $\mathbf{e}_i = (0, \ldots, 0, 1_i, 0, \ldots, 0)$ is the unit vector.

Let us now define the vectors

$$|i, t\rangle = V(t)^\dagger |\mathbf{e}_i\rangle, \qquad i = 1, \ldots, k. \tag{1.107}$$

Using (1.105) we find that these vectors satisfy the following evolution equation in Hilbert space:

$$\frac{d}{dt}|i, t\rangle = M_{\mathrm{H}}^\dagger(t)|i, t\rangle, \qquad |i, 0\rangle = |\mathbf{e}_i\rangle, \qquad i = 1, \ldots, k, \tag{1.108}$$

where $M_{\mathrm{H}}(t) = V(t)^{-1}M(t)V(t)$ is the "Hamiltonian" in "Heisenberg picture" (see section 1.2). Taking into account (1.104) and (1.106) we obtain the following formula relating the solution to the system (1.102) and the solution of (1.108):

$$x_i(\mathbf{x}_0, t) = \langle i, t | \mathbf{x}_0\rangle e^{\frac{1}{2}\mathbf{x}_0^2}, \qquad i = 1, \ldots, k. \tag{1.109}$$

We have thus shown that the problem of integrating the nonlinear dynamical system (1.102) is equivalent to solving the system of linear, abstract evolution equations in

Hilbert space (1.108). On writing (1.108) in the occupation number representation we arrive at the following system of differential-difference equations:

$$\frac{dx_{i\mathbf{n}}}{dt} = \sum_{\mathbf{n}'\in \mathbf{Z}_+^k} M^\dagger_{\mathrm{H}\mathbf{n}\mathbf{n}'}(t)x_{i\mathbf{n}'}, \tag{1.110a}$$

$$x_{i\mathbf{n}}(0) = \delta_{n_i1}\prod_{j\neq i}\delta_{n_j0}, \qquad i = 1, \ldots, k, \tag{1.110b}$$

where $x_{i\mathbf{n}}(t) = \langle \mathbf{n}|i,t\rangle$ and $M^\dagger_{\mathrm{H}\mathbf{n}\mathbf{n}'}(t) = \langle \mathbf{n}|M^\dagger_{\mathrm{H}}(t)|\mathbf{n}'\rangle$.

As an immediate consequence of (1.109), (C.10) and (C.25), we find that the solution of the system (1.102) is linked to the solution of (1.110) by

$$x_i(\mathbf{x}_0,t) = \sum_{\mathbf{n}\in\mathbf{Z}_+^k} x_{i\mathbf{n}}(t)\prod_{j=1}^k \frac{x_{0j}^{n_j}}{\sqrt{n_j!}}, \qquad i = 1, \ldots, k.$$

It thus appears that the canonical Hilbert space formalism amounts a far-reaching generalization of the Wackerbauer approach. On the one hand, we linearize the general nonautonomous systems (1.102) including the autonomous systems with polynomial nonlinearity considered by Wackerbauer as the special case. On the other hand, the Wackerbauer technique corresponds to the particular occupation number representation in the presented Hilbert space approach. It should also be noted that the canonical formalism presented herein enables one to determine in a simple way the matrix of the infinite linear system satisfied by the modes $x_{i\mathbf{n}}$.

REMARK 1. Consider the real autonomous system

$$\frac{d\mathbf{x}}{dt} = \mathbf{F}(\mathbf{x}), \qquad \mathbf{x}(0) = \mathbf{x}_0, \tag{1.111}$$

where $\mathbf{F}$: $\mathbf{R}^k \to \mathbf{R}^k$ is analytic in $\mathbf{x}$. The corresponding equation (1.108) takes then the form

$$\frac{d}{dt}|i,t\rangle = M^\dagger|i,t\rangle, \qquad |i,0\rangle = |\mathbf{e}_i\rangle, \qquad i = 1, \ldots, k, \tag{1.112}$$

where $M^\dagger = \mathbf{F}(\mathbf{a}^\dagger)\cdot\mathbf{a}$. Writing this equation in the Bargmann representation (see appendix C) we obtain the partial differential equation such that

$$\frac{\partial}{\partial t}\tilde{\phi}_i(\mathbf{x},t) = \mathbf{F}(\mathbf{x})\cdot\frac{\partial}{\partial \mathbf{x}}\tilde{\phi}_i(\mathbf{x},t), \qquad \tilde{\phi}_i(\mathbf{x},0) = x_i, \qquad i = 1, \ldots, k, \tag{1.113}$$

where $\tilde{\phi}_i(\mathbf{x},t) = \langle \mathbf{x}|i,t\rangle e^{\frac{1}{2}\mathbf{x}^2}$. We have thus shown that (1.112) is the "quantal" basis-independent generalization of the Liouville equation [23] (1.113) corresponding to the system (1.111). ■

REMARK 2. Taking into account (1.107) and (C.7) as well as

$$V(t)^\dagger|\mathbf{0}\rangle = |\mathbf{0}\rangle$$

which follows directly from (1.30) (see also (B.9a)), we find

$$|i,t\rangle = a_i(t)^\dagger|\mathbf{0}\rangle = x_i(\mathbf{a}^\dagger,t)|\mathbf{0}\rangle, \qquad i = 1, \ldots, k,$$

where $a_i(t)$ are the time-dependent Bose annihilation operators discussed in section 1.2 (see (1.36)) and $\mathbf{x}(\mathbf{x}_0,t)$ is the solution of (1.102). Thus, it turns out that solutions to (1.102) are the symbols of the vectors (1.107) (see appendix C). ■

### *1.4.2 Koopman-Alanson linearization*

In this section we show that both the classical Koopman linearization and the generalization of this approach which has been recently introduced by Alanson [9] amount a particular case of the general technique of reduction of nonlinear dynamical systems to linear, abstract evolution equations in Hilbert space [24]. This technique was used by the author for formulation of the Hilbert space formalism described in section 1.1. We begin by recalling the Koopman linearization (compare the last chapter of the book [25]). Consider the Hamiltonian system

$$\begin{aligned}\frac{d\mathbf{q}}{dt} &= \frac{\partial H}{\partial \mathbf{p}},\\ \frac{d\mathbf{p}}{dt} &= -\frac{\partial H}{\partial \mathbf{q}}, \qquad \mathbf{q}(0) = \mathbf{q}_0,\ \mathbf{p}(0) = \mathbf{p}_0.\end{aligned} \tag{1.114}$$

We define a differential operator $U(t)$ acting on complex-valued functions on the phase space $\mathbf{R}^{2k}$ of the system (1.114) as follows:

$$(U(t)\phi)(\mathbf{q},\mathbf{p}) = \phi(g^t(\mathbf{q},\mathbf{p})), \tag{1.115}$$

where $\phi$: $\mathbf{R}^{2k} \to \mathbf{C}$ and $g^t$ is the (local) phase flow, that is $g^t(\mathbf{q}_0,\mathbf{p}_0)$ is the solution of (1.114). On differentiating both sides of (1.115) with respect to time we find that the function $\phi(\mathbf{q},\mathbf{p},t)$ defined as

$$\phi(\mathbf{q},\mathbf{p},t) = (U(t)\phi)(\mathbf{q},\mathbf{p})$$

satisfies the following evolution equation:

$$\frac{\partial}{\partial t}\phi(\mathbf{q},\mathbf{p},t) = X_H(\mathbf{q},\mathbf{p})\phi(\mathbf{q},\mathbf{p},t), \qquad \phi(\mathbf{q},\mathbf{p},0) = \phi(\mathbf{q},\mathbf{p}), \tag{1.116}$$

where $X_H(\mathbf{q},\mathbf{p})$ is the vector field corresponding to the system (1.114)

$$X_H(\mathbf{q},\mathbf{p}) = \frac{\partial H}{\partial \mathbf{p}}\cdot\frac{\partial}{\partial \mathbf{q}} - \frac{\partial H}{\partial \mathbf{q}}\cdot\frac{\partial}{\partial \mathbf{p}}. \tag{1.117}$$

Now, an easy integration by parts shows that the differential operator (1.117) is skew-Hermitian with respect to the inner product

$$\langle\phi|\psi\rangle = \int d^kp\,d^kq\,\phi^*(\mathbf{q},\mathbf{p})\psi(\mathbf{q},\mathbf{p}).$$

Consequently, (1.116) can be rewritten as the Schrödinger equation in the Hilbert space $L^2(\mathbf{R}^{2k}, d^kpd^kq)$

$$i\frac{\partial}{\partial t}\phi(\mathbf{q},\mathbf{p},t) = \mathcal{H}(\mathbf{q},\mathbf{p})\phi(\mathbf{q},\mathbf{p},t), \tag{1.118}$$

where the Hermitian Hamiltonian $\mathcal{H}(\mathbf{q},\mathbf{p})$ is of the form

$$\mathcal{H}(\mathbf{q},\mathbf{p}) = i\left(\frac{\partial H}{\partial \mathbf{p}}\cdot\frac{\partial}{\partial \mathbf{q}} - \frac{\partial H}{\partial \mathbf{q}}\cdot\frac{\partial}{\partial \mathbf{p}}\right).$$

Moreover, taking into account (1.118) we find that the operator $U(t)$ given by (1.115) and the "quantal" Hamiltonian $\hat{H}$ such that $(\hat{H}\phi)(\mathbf{q},\mathbf{p}) = \mathcal{H}(\mathbf{q},\mathbf{p})\phi(\mathbf{q},\mathbf{p})$ are related by

$$U(t) = e^{i\hat{H}t}.$$

Therefore, $U(t)$ is the unitary map (evolution operator) satisfying the Schrödinger equation

$$i\frac{dU}{dt} = \hat{H}U, \qquad U(0) = I.$$

It thus appears that to each classical Hamiltonian system (1.114) there corresponds the Schrödinger equation (1.118) in the Hilbert space $L^2(\mathbf{R}^{2k}, d^kpd^kq)$. Due to the fact that the measure $d^kpd^kq$ is invariant under the phase flow $g^t$, the Koopman linearization has become a powerful tool in the ergodic theory (see remark 2 on page 26).

We now proceed to describe the Alanson approach. Consider the real autonomous system

$$\frac{d\mathbf{q}}{dt} = \mathbf{F}(\mathbf{q}), \qquad \mathbf{q}(0) = \mathbf{q}_0, \tag{1.119}$$

where $\mathbf{F}$: $\mathbf{R}^k \to \mathbf{R}^k$ is a differentiable vector field. Let us introduce a differential operator $U(t)$ acting on complex-valued functions on the phase space $\mathbf{R}^k$ of the system (1.119) such that

$$\begin{aligned}(U(t)\phi)(\mathbf{q}) &= \left[\det\left(\frac{\partial g^{-t}\mathbf{q}}{\partial \mathbf{q}}\right)\right]^{\frac{1}{2}}\phi(g^{-t}\mathbf{q}) \\ &= \exp\left[\frac{1}{2}\int_0^{-t} d\tau\,(\mathrm{div}\mathbf{F})(g^{\tau}\mathbf{q})\right]\phi(g^{-t}\mathbf{q}),\end{aligned} \tag{1.120}$$

where $\phi$: $\mathbf{R}^k \to \mathbf{C}$, $g^t$ is the phase flow, i.e. $g^t\mathbf{q}_0$ is the solution of (1.119) and $\det\left(\frac{\partial g^t\mathbf{q}}{\partial \mathbf{q}}\right) = \exp\left[\int_0^t d\tau\,(\mathrm{div}\mathbf{F})(g^{\tau}\mathbf{q})\right]$, with $\mathrm{div}\mathbf{F} = \frac{\partial}{\partial \mathbf{q}}\cdot\mathbf{F}(\mathbf{q})$, is the Jacobian. An easy differentiation shows that the function

$$\phi(\mathbf{q},t) = (U(t)\phi)(\mathbf{q})$$

satisfies the evolution equation of the form

$$\frac{\partial}{\partial t}\phi(\mathbf{q},t) = -[X_{\mathbf{F}}(\mathbf{q}) + \frac{1}{2}(\mathrm{div}\mathbf{F})(\mathbf{q})]\phi(\mathbf{q},t) \tag{1.121a}$$

$$= -\frac{1}{2}\left[\mathbf{F}(\mathbf{q})\cdot\frac{\partial}{\partial\mathbf{q}} + \frac{\partial}{\partial\mathbf{q}}\cdot\mathbf{F}(\mathbf{q})\right]\phi(\mathbf{q},t), \qquad \phi(\mathbf{q},0) = \phi(\mathbf{q}), \tag{1.121b}$$

where $X_{\mathbf{F}} = \mathbf{F}(\mathbf{q})\cdot\frac{\partial}{\partial\mathbf{q}}$ is the vector field corresponding to (1.119).

It is easy to verify that the differential operator in the square brackets from the right-hand side of (1.121b) is skew-Hermitian in the Hilbert space $L^2(\mathbf{R}^k, d^kq)$ specified by the inner product

$$\langle\phi|\psi\rangle = \int d^kq\,\phi^*(\mathbf{q})\psi(\mathbf{q}).$$

Therefore, eq. (1.121) can be written as the Schrödinger equation in the Hilbert space $L^2(\mathbf{R}^k, d^kq)$

$$i\frac{\partial}{\partial t}\phi(\mathbf{q},t) = \mathcal{H}(\mathbf{q})\phi(\mathbf{q},t), \qquad \phi(\mathbf{q},0) = \phi(\mathbf{q}), \tag{1.122}$$

where the Hamiltonian $\mathcal{H}(\mathbf{q})$ is

$$\mathcal{H}(\mathbf{q}) = -\frac{i}{2}\left[\mathbf{F}(\mathbf{q})\cdot\frac{\partial}{\partial\mathbf{q}} + \frac{\partial}{\partial\mathbf{q}}\cdot\mathbf{F}(\mathbf{q})\right]. \tag{1.123}$$

Thus, it turns out that one can associate with classical dynamical systems (1.119) the Schrödinger equation (1.122). Comparing the two sides of (1.115), (1.116) and (1.120), (1.121a), respectively and taking into account that the vector fields of Hamiltonian systems are divergenceless, we find that the Alanson approach amounts a generalization of the Koopman linearization to the case of the autonomous systems (1.119). Nevertheless, the Koopman work [1] was not referenced in [9]. It should also be noted that the measure $d^kq$ is not preserved under the phase flow $g^t$ unless $\mathrm{div}\mathbf{F} = 0$ (see remark 2 below). Thus, on the contrary to the Koopman linearization, the eventual applications of the Alanson's observations in the ergodic theory seem to be limited.

REMARK 1. The formula (1.120) is a special case of the general relation [26]:

$$(U(g)\phi)(x) = [r_g(g^{-1}x)]^{\frac{1}{2}}\phi(g^{-1}x), \qquad x \in X, \tag{1.124}$$

defining the unitary representation $U(g)$ of a group $G$ acting on the Borel space $X$, in $L^2(X, d\mu)$, where the measure $\mu$ need not be invariant, that is $\mu(A) = \mu(g^{-1}A)$ need not hold, where $A$ is a Borel subset of $X$. Here, $r_g(x)$ is a positive Borel function which is a version of the Radon-Nikodym derivative: $r_g(x) = \frac{d\mu(x)}{d\mu(gx)}$. Indeed, it is easy to verify that if $X$ is the phase space of the system (1.119), $G = \mathbf{R}$ and the action of $G$ on $X$ is given by the dynamical group of the system, i.e. by the phase flow, then

(1.124) reduces to (1.120). Notice that the case with invariant measure $d\mu$, when (1.124) takes the form

$$(U(g)\phi)(x) = \phi(g^{-1}x), \qquad x \in X,$$

amounts a generalization of the transformation (1.115) considered by Koopman to the case of an arbitrary $G$-space $X$. ■

REMARK 2. We first recall the classical Liouville theorem on the invariant measure [2]. Consider the nonlinear dynamical system

$$\frac{d\mathbf{x}}{dt} = \mathbf{F}(\mathbf{x}), \qquad \mathbf{x}(0) = \mathbf{x}_0, \tag{1.125}$$

where $\mathbf{F}$ is a smooth vector field on a smooth $k$-dimensional manifold $\mathbf{M}$.

The measure $\mu$ such that $d\mu = \rho(\mathbf{x})d^k x$, where $\rho(\mathbf{x})$ is a density, is said to be invariant under the flow $g^t$, where $g^t\mathbf{x}_0$ is the solution to (1.125), if the condition

$$\int\limits_{\mathbf{M}} \phi\rho\, d^k x = \int\limits_{\mathbf{M}} \phi(g^t\mathbf{x})\rho\, d^k x, \qquad |t| < t_0,$$

is valid for every smooth function $\phi$ and $t_0$ such that the closure of supp$\phi$ and closure of supp$\phi(g^t\mathbf{x})$ for every $|t| < t_0$, are contained in a chart $U$.

According to the Liouville theorem the measure $\mu$ such that $d\mu = \rho(\mathbf{x})d^k x$, where $\rho(\mathbf{x})$ is a density, is invariant under the flow $g^t$, where $g^t\mathbf{x}_0$ is the solution of (1.125), if and only if

$$\operatorname{div}\rho\mathbf{F} = 0.$$

By virtue of the $L^2$-version of the Liouville theorem [27] (see also [2]) the operator $iX_{\mathbf{F}}$, where $X_{\mathbf{F}} = \mathbf{F}(\mathbf{x})\cdot\frac{\partial}{\partial\mathbf{x}}$ is the vector field corresponding to (1.125), defined on some dense subspace of $L^2(\mathbf{M}, d\mu)$, is symmetric whenever $d\mu = \rho(\mathbf{x})d^k x$ is an invariant measure. Using the Koopman approach it follows that then the map

$$(U(t)\phi)(\mathbf{x}) = \phi(g^t\mathbf{x}) \tag{1.126}$$

is unitary one on $L^2(\mathbf{M}, d\mu)$.

We recall that the fundamental object in the ergodic theory is a triple $(\mathbf{M}, \mu, g^t)$ called dynamical system, where $\mu$ is an invariant measure under the phase flow $g^t$ on $\mathbf{M}$. The observations of Koopman enable one to study the properties of such dynamical systems by means of unitary maps on $L^2(\mathbf{M}, d\mu)$ defined by (1.126). ■

REMARK 3. Following Alanson we note that the Hamiltonian (1.123) can be written as

$$H = \frac{1}{2}(\mathbf{F}(\hat{\mathbf{q}})\cdot\hat{\mathbf{p}} + \hat{\mathbf{p}}\cdot\mathbf{F}(\hat{\mathbf{q}})), \tag{1.127}$$

where $(H\phi)(\mathbf{p},\mathbf{q}) = \mathcal{H}(\mathbf{p},\mathbf{q})\phi(\mathbf{p},\mathbf{q})$ and $\hat{\mathbf{q}}$, $\hat{\mathbf{p}}$ are realizations of the position and momentum operators, respectively in the coordinate representation (see appendix D), that is,

$$(\hat{\mathbf{q}}\phi)(\mathbf{q}) = \mathbf{q}\phi(\mathbf{q}), \qquad (\hat{\mathbf{p}}\phi)(\mathbf{q}) = -i\frac{\partial}{\partial \mathbf{q}}\phi(\mathbf{q}).$$

The Heisenberg equations of motion satisfied by the time-dependent position and momentum operators (see appendix B) are of the form

$$\frac{d\hat{\mathbf{q}}(t)}{dt} = -i[\hat{\mathbf{q}}(t), H_{\rm H}] = \mathbf{F}(\hat{\mathbf{q}}(t)), \tag{1.128a}$$

$$\frac{d\hat{\mathbf{p}}(t)}{dt} = -i[\hat{\mathbf{p}}(t), H_{\rm H}] = -\frac{1}{2}\sum_{i=1}^{k}\left(\frac{\partial F_i(\hat{\mathbf{q}}(t))}{\partial \hat{\mathbf{q}}(t)}\hat{p}_i(t) + \hat{p}_i(t)\frac{\partial F_i(\hat{\mathbf{q}}(t))}{\partial \hat{\mathbf{q}}(t)}\right), \tag{1.128b}$$

where $H_{\rm H} = \frac{1}{2}[\mathbf{F}(\hat{\mathbf{q}}(t))\cdot\hat{\mathbf{p}}(t) + \hat{\mathbf{p}}(t)\cdot\mathbf{F}(\hat{\mathbf{q}}(t))]$ is the Hamiltonian in Heisenberg picture (since the Hamiltonian in Schrödinger picture (1.127) does not depend on time, therefore we have $H = H_{\rm H}$ (see appendix B)). The system of evolution operator equations (1.128) is the "quantized" version of the following classical dynamical system:

$$\frac{d\mathbf{q}}{dt} = \mathbf{F}(\mathbf{q}), \tag{1.129a}$$

$$\frac{d\mathbf{p}}{dt} = -\sum_{j=1}^{k} p_j \frac{\partial F_j}{\partial \mathbf{q}}, \tag{1.129b}$$

where we apply the usual Jordan rules of quantization relying on substitutions $\mathbf{q} \to \hat{\mathbf{q}}$, $\mathbf{p} \to \hat{\mathbf{p}} = -i\frac{\partial}{\partial \mathbf{q}}$, as well as the symmetrization $f(\mathbf{q})g(\mathbf{p}) \to \frac{1}{2}[f(\hat{\mathbf{q}})g(\hat{\mathbf{p}}) + g(\hat{\mathbf{p}})f(\hat{\mathbf{q}})]$, where $f$, $g$ are real analytic functions, to insure the Hermitian property. Evidently, the system (1.129) is Hamilton one with the Hamiltonian such that

$$H(\mathbf{p},\mathbf{q}) = \mathbf{p}\cdot\mathbf{F}(\mathbf{q}).$$

Such method for reduction of the system (1.119) to the Hamiltonian system (1.129) is usually known as the *method of doubling variables* [28]. As far as we are aware this method was invented in 1956 by Pontrjagin who applied this in formulation of his maximum principle playing an important role in the theory of optimal processes. Notice that equations analogous to (1.128) are formally satisfied by the time-dependent Bose operators $\mathbf{a}(t)$, $\mathbf{a}^\dagger(t)$ (see section 1.2). Namely, we have

$$\frac{d\mathbf{a}(t)}{dt} = [\mathbf{a}(t), M_{\rm H}] = \mathbf{F}(\mathbf{a}(t)), \tag{1.130a}$$

$$\frac{d\mathbf{a}^\dagger(t)}{dt} = [\mathbf{a}^\dagger(t), M_{\rm H}] = -\sum_{i=1}^{k} a_i^\dagger(t)\frac{\partial F_i(\mathbf{a}(t))}{\partial \mathbf{a}(t)}, \tag{1.130b}$$

where $M_{\rm H} = M = \mathbf{a}^\dagger(t)\cdot\mathbf{F}(\mathbf{a}(t))$ is the "Hamiltonian" corresponding to (1.129a). As is easily seen from the last equation of (1.130b), the system (1.130) cannot be treated as a result of any nonambiguous "quantization" of (1.129). Thus the adequate Hilbert space counterpart of the method of doubling coordinates is the Heisenberg picture within the Alanson approach. Finally, we note that the observations presented above can be immediately generalized to involve the case of nonautonomous systems (1.119) with the right-hand side dependent explicitly on $t$. ■

We now introduce a general technique for reduction of classical dynamical systems to linear, abstract evolution equations in Hilbert space and we show that it includes the Alanson approach and thus the Koopman linearization as a special case. Consider the nonlinear dynamical system (real or complex)

$$\frac{d\mathbf{x}}{dt} = \mathbf{F}(\mathbf{x}), \qquad \mathbf{x}(0) = \mathbf{x}_0, \tag{1.131}$$

where $\mathbf{F}\colon \mathbf{R}^k \to \mathbf{R}^k$ is analytic in $\mathbf{x}$.

We recall that the Carleman ansatz for (1.131) is given by

$$x_{\mathbf{n}} = \prod_{i=1}^{k} (x_i(t))^{n_i} \tag{1.132}$$

The linear differential-difference equation implied by (1.132) can be written as

$$\frac{dx_{\mathbf{n}}}{dt} = \sum_{\mathbf{n}' \in \mathbf{Z}_+^k} M_{\mathbf{n}\mathbf{n}'} x_{\mathbf{n}'}. \tag{1.133}$$

As recognized by Carleman [3], the linear system of the form (1.133) can be obtained with the help of the more general ansatz

$$x_{\mathbf{n}} = p_{\mathbf{n}}(\mathbf{x}) \exp(\phi(\mathbf{x})), \tag{1.134}$$

where $p_{\mathbf{n}}$, $\phi$ are polynomial in $\mathbf{x}$, and

$$\int_{\Omega} d\nu(\mathbf{x})\, w(\mathbf{x}) p_{\mathbf{n}}(\mathbf{x}) p_{\mathbf{m}}(\mathbf{x}) = \delta_{\mathbf{n}\mathbf{m}},$$

where $\delta_{\mathbf{nm}} = \prod_{i=1}^{k} \delta_{n_i m_i}$, that is $p_{\mathbf{n}}$ is a multidimensional generalization of an orthogonal polynomial; $w(\mathbf{x})$ is a corresponding weight function.

Based on the observations of section 1.1 we now describe the scheme of converting (1.131) into the linear, abstract evolution equation in Hilbert space such that the realization of this abstract equation in the occupation number representation corresponds to the generalized ansatz of the form (1.134). Consider the ansatz (1.134). We seek the commuting operators $A_i$, $i = 1, \ldots, k$, in a Hilbert space, satisfying

$$\mathbf{A}|\mathbf{x}\rangle = \mathbf{x}|\mathbf{x}\rangle, \tag{1.135}$$

where the common eigenvectors $|\mathbf{x}\rangle$, with $\mathbf{x} \in \mathbf{R}^k$ or $\mathbf{x} \in \mathbf{C}^k$, of operators $A_i$, are related to the ansatz (1.134) by

$$\langle \mathbf{n}|\mathbf{x}(t)\rangle = x_{\mathbf{n}}, \tag{1.136}$$

where the vectors $|\mathbf{n}\rangle$ span the occupation number representation and $\mathbf{x}(t)$ fulfils (1.131).

Taking into account (1.136) and (1.133) we find that the vectors $|\mathbf{x}(t)\rangle$ obey the linear evolution equation in Hilbert space of the form

$$\frac{d}{dt}|\mathbf{x}(t)\rangle = M|\mathbf{x}(t)\rangle, \qquad |\mathbf{x}(0)\rangle = |\mathbf{x}_0\rangle, \tag{1.137}$$

where the matrix elements of the "Hamiltonian" $M$

$$\langle \mathbf{n}|M|\mathbf{n}'\rangle = M_{\mathbf{n}\mathbf{n}'},$$

coincide with the embedding matrix $M_{\mathbf{n}\mathbf{n}'}$ from the right-hand side of eq. (1.133) resulting from the ansatz (1.134).

Evidently, the solution $\mathbf{x}(\mathbf{x}_0, t)$ to (1.131) is linked to the solution $|\mathbf{x}_0, t\rangle$ of (1.137) by

$$\mathbf{A}|\mathbf{x}_0, t\rangle = \mathbf{x}(\mathbf{x}_0, t)|\mathbf{x}_0, t\rangle. \tag{1.138}$$

Differentiating both sides of (1.138) we arrive at the following formula relating operators $\mathbf{A}$ and the "Hamiltonian" $M$:

$$[\mathbf{A}, M] = \mathbf{F}(\mathbf{A}). \tag{1.139}$$

Thus whenever we success in finding operators $\mathbf{A}$ and $M$ satisfying (1.139), (1.135) and (1.136), then the nonlinear dynamical system (1.131) can be cast into the linear, abstract evolution equation in Hilbert space (1.137). As mentioned in the introduction to this section, the algorithm described above was used by the author for formulation of the Hilbert space formalism described in section 1.1. Indeed, consider the system (complex or real)

$$\frac{d\mathbf{z}}{dt} = \mathbf{F}(\mathbf{z}), \qquad \mathbf{z}(0) = \mathbf{z}_0, \tag{1.140}$$

where $\mathbf{F}$: $\mathbf{C}^k \to \mathbf{C}^k$ is analytic in $\mathbf{z}$.

The following solution of (1.139) can be easily found

$$\mathbf{A} = \mathbf{a}, \qquad M = \mathbf{a}^\dagger \cdot \mathbf{F}(\mathbf{a}), \tag{1.141}$$

where $\mathbf{a}$, $\mathbf{a}^\dagger$ are the Bose annihilation and creation operators, respectively.

The eigenvectors of the annihilation operators $\mathbf{a}$ are the coherent states. This means that the counterpart of (1.135) takes the form

$$\mathbf{a}|\mathbf{z}\rangle = \mathbf{z}|\mathbf{z}\rangle. \tag{1.142}$$

We define the time-dependent coherent states $|\mathbf{z}(t)\rangle$, where $\mathbf{z}(t)$ satisfies (1.140) as follows:

$$|\mathbf{z}(t)\rangle = \exp(-\tfrac{1}{2}|\mathbf{z}|^2)\exp(\mathbf{z}\cdot\mathbf{a}^\dagger)|\mathbf{0}\rangle. \tag{1.143}$$

These vectors obey the relation of the form (1.136). Namely, we have

$$z_{\mathbf{n}}(t) = \langle \mathbf{n}|\mathbf{z}(t)\rangle = \left(\prod_{i=1}^{k} \frac{(z_i(t))^{n_i}}{\sqrt{n_i!}}\right)\exp(-\tfrac{1}{2}|\mathbf{z}_0|^2) = p_{\mathbf{n}}(\mathbf{z}(t))\exp(-\tfrac{1}{2}|\mathbf{z}_0|^2),$$

where the polynomials $p_{\mathbf{n}}(\mathbf{z}^*)$ are orthonormal in the Bargmann space of entire functions specified by the inner product (C.33) (see (C.36)).

The vectors (1.143) fulfil the following abstract evolution equation in Hilbert space:

$$\frac{d}{dt}|\mathbf{z}(t)\rangle = M|\mathbf{z}(t)\rangle, \qquad |\mathbf{z}(0)\rangle = |\mathbf{z}_0\rangle. \tag{1.144}$$

The solution $\mathbf{z}(\mathbf{z}_0, t)$ of the system (1.140) is obviously related to the solution $|\mathbf{z}_0, t\rangle$ of (1.144) by

$$\mathbf{a}|\mathbf{z}_0, t\rangle = \mathbf{z}(\mathbf{z}_0, t)|\mathbf{z}_0, t\rangle.$$

We have thus rediscovered the Hilbert space formalism described in section 1.1.

We now derive an alternative solution to (1.139), (1.135) and (1.136) corresponding to the Alanson linearization. Consider the real analytic system

$$\frac{d\mathbf{q}}{dt} = \mathbf{F}(\mathbf{q}), \qquad \mathbf{q}(0) = \mathbf{q}_0, \tag{1.145}$$

where $\mathbf{F}\colon \mathbf{R}^k \to \mathbf{R}^k$ is analytic in $\mathbf{q}$.

The following operators are easily seen to satisfy (1.139):

$$\mathbf{A} = \hat{\mathbf{q}}, \qquad M = -i\hat{\mathbf{p}}\cdot\mathbf{F}(\hat{\mathbf{q}}), \tag{1.146}$$

where $\hat{\mathbf{q}}$ and $\hat{\mathbf{p}}$ are the position and momentum operators, respectively (see appendix D).

The common eigenvectors $|\mathbf{q}\rangle$ of the position operators, given by

$$\hat{\mathbf{q}}|\mathbf{q}\rangle = \mathbf{q}|\mathbf{q}\rangle, \tag{1.147}$$

can be introduced as

$$|\mathbf{q}\rangle = \pi^{-\frac{k}{4}}\exp(\tfrac{1}{2}\mathbf{q}^2)\exp[-\tfrac{1}{2}(\mathbf{a}^\dagger - \sqrt{2}\,\mathbf{q})^2]|\mathbf{0}\rangle. \tag{1.148}$$

It is clear that (1.147) is the counterpart of (1.135). Furthermore, the projection of the vectors (1.148), where $\mathbf{q}$ satisfies (1.145), onto the basis vectors of the occupation number representation gives

$$q_{\mathbf{n}} = \langle \mathbf{n}|\mathbf{q}\rangle = \left(\prod_{i=1}^{k} \bar{H}_{n_i}(q_i)\right)\exp(-\tfrac{1}{2}\mathbf{q}^2) = p_{\mathbf{n}}(\mathbf{q})\exp(-\tfrac{1}{2}\mathbf{q}^2), \tag{1.149}$$

where $\bar{H}_{n_i}$ are normalized Hermite polynomials.

The fact that the polynomials $p_{\mathbf{n}}(\mathbf{q})$ are orthonormal is straightforward. We recall (see (D.4)) that the following relation holds true:

$$\int_{\mathbf{R}^k} d^k q \exp(-\mathbf{q}^2)\, p_{\mathbf{n}}(\mathbf{q}) p_{\mathbf{m}}(\mathbf{q}) = \delta_{\mathbf{nm}}.$$

Therefore, (1.149) is the counterpart of (1.136). An easy inspection based on (D.2b) shows that the vectors $|\mathbf{q}(t)\rangle$ obey the abstract evolution equation in Hilbert space (1.137) such that

$$\frac{d}{dt}|\mathbf{q}(t)\rangle = M|\mathbf{q}(t)\rangle, \qquad |\mathbf{q}(0)\rangle = |\mathbf{q}_0\rangle, \tag{1.150}$$

where $M = -i\hat{\mathbf{p}}\cdot\mathbf{F}(\hat{\mathbf{q}})$.

Evidently, the solution $\mathbf{q}(\mathbf{q}_0, t)$ of the system (1.145) is linked to the solution $|\mathbf{q}_0, t\rangle$ of (1.150) by

$$\hat{\mathbf{q}}|\mathbf{q}_0, t\rangle = \mathbf{q}(\mathbf{q}_0, t)|\mathbf{q}_0, t\rangle.$$

We have thus shown that the nonlinear dynamical system (1.145) can be brought down to the linear, abstract evolution equation in Hilbert space (1.150).

We now demonstrate that (1.150) can be furthermore reduced to the Schrödinger equation. Let us introduce the vectors

$$|\mathbf{q}(t)\rangle' = \exp\left(\frac{1}{2}\int_0^t d\tau \,\mathrm{div}\mathbf{F}(\mathbf{q}(\tau))\right)|\mathbf{q}(t)\rangle, \tag{1.151}$$

where $|\mathbf{q}(t)\rangle$ fulfils (1.150). Taking into account (D.3a) we find that the vectors (1.151) satisfy the following Schrödinger equation:

$$i\frac{d}{dt}|\mathbf{q}(t)\rangle' = H|\mathbf{q}(t)\rangle', \qquad |\mathbf{q}(0)\rangle' = |\mathbf{q}_0\rangle, \tag{1.152}$$

where the Hamiltonian $H$ is the symmetrization of the operator $iM$, i.e.,

$$H = \frac{1}{2}(\hat{\mathbf{p}}\cdot\mathbf{F}(\hat{\mathbf{q}}) + \mathbf{F}(\hat{\mathbf{q}})\cdot\hat{\mathbf{p}}).$$

An immediate consequence of (1.147) and (1.151) is the following eigenvalue equation relating the solution $\mathbf{q}(\mathbf{q}_0, t)$ of the system (1.145) and the solution $|\mathbf{q}_0, t\rangle'$ of (1.152):

$$\hat{\mathbf{q}}|\mathbf{q}_0, t\rangle' = \mathbf{q}(\mathbf{q}_0, t)|\mathbf{q}_0, t\rangle'.$$

Thus, it turns out that the nonlinear dynamical system (1.145) can be cast into the abstract Schrödinger equation (1.152). By taking the Hermitian conjugate of (1.152) and projecting the obtained equation onto the abstract state $|\phi\rangle$ we arrive at the following equation:

$$i\frac{\partial}{\partial t}\phi(\mathbf{q}) = \mathcal{H}(\mathbf{q})\phi(\mathbf{q}), \qquad \phi(\mathbf{q}(0)) = \phi(\mathbf{q}_0), \tag{1.153}$$

where

$$\phi(\mathbf{q}) = {}'\langle \mathbf{q}(-t)|\phi\rangle,$$

and

$$\mathcal{H}(\mathbf{q}) = -\frac{i}{2}\left(\frac{\partial}{\partial \mathbf{q}}\cdot\mathbf{F}(\mathbf{q}) + \mathbf{F}(\mathbf{q})\cdot\frac{\partial}{\partial \mathbf{q}}\right).$$

Now treating the vectors $|\mathbf{q}(-t)\rangle'$ as a moving basis of the coordinate representation we find that (1.153) can be regarded as the projection of the abstract Schrödinger equation

$$i\frac{d}{dt}|\phi(t)\rangle = H|\phi(t)\rangle, \qquad |\phi(0)\rangle = |\phi\rangle,$$

onto the vector $|\mathbf{q}_0\rangle$ such that

$$|\mathbf{q}(-t)\rangle' = e^{-itH}|\mathbf{q}_0\rangle.$$

In other words, (1.153) can be written in the following equivalent form:

$$i\frac{\partial}{\partial t}\tilde{\phi}(\mathbf{q}_0, t) = \mathcal{H}(\mathbf{q}_0)\tilde{\phi}(\mathbf{q}_0, t), \qquad \tilde{\phi}(\mathbf{q}_0, 0) = \tilde{\phi}(\mathbf{q}_0), \tag{1.154}$$

where

$$\tilde{\phi}(\mathbf{q}_0, t) = \langle \mathbf{q}_0|\phi(t)\rangle \equiv \phi(\mathbf{q}(\mathbf{q}_0, t)),$$

and

$$\mathcal{H}(\mathbf{q}_0) = -\frac{i}{2}\left(\frac{\partial}{\partial \mathbf{q}_0}\cdot\mathbf{F}(\mathbf{q}_0) + \mathbf{F}(\mathbf{q}_0)\cdot\frac{\partial}{\partial \mathbf{q}_0}\right).$$

Evidently, (1.154) coincides with the Schrödinger equation (1.122) derived by Alanson. We have thus demonstrated that the Koopman-Alanson linearization is included by the general method for reduction of nonlinear dynamical systems to linear, abstract evolution equations in Hilbert space described above as a particular case.

REMARK 1. The reader may notice that the above algorithm of converting (1.145) into the abstract evolution equation in Hilbert space can be immediately applied to the case with the solution of (1.139) of the form

$$\mathbf{A} = \hat{\mathbf{p}}, \qquad M = i\hat{\mathbf{q}}\cdot\mathbf{F}(\hat{\mathbf{p}}).$$

Nevertheless, in view of (D.8), a picture is then merely obtained which is unitarily equivalent to that one resulting from (1.146). ■

REMARK 2. Note that the method for reduction of nonlinear dynamical systems to abstract evolution equations in Hilbert space described above can be immediately generalized to include nonautonomous systems. It should also be noted that the symmetrization of the operator $iM$, where $M$ is given by (1.141), contains in the case with nonlinear systems (1.140) the terms nonlinear in $\mathbf{a}^\dagger$. The presence of such terms was shown in ref. [29] to lead to violation of the condition (1.142) during the time evolution (the coherent states become unstable). It thus appears that (1.144) cannot be reduced to the Schrödinger equation. ■

We end this section with discussion of the transformations like (1.115) in the context of the Hilbert space approach introduced in this book. Consider the real autonomous system

$$\frac{d\mathbf{x}}{dt} = \mathbf{F}(\mathbf{x}), \qquad \mathbf{x}(0) = \mathbf{x}_0, \tag{1.155}$$

where $\mathbf{F}\colon \mathbf{R}^k \to \mathbf{R}^k$ is analytic in $\mathbf{x}$.

Let us introduce the following differential operator acting on real or complex valued functions defined on the phase space of the system (1.155):

$$(U(t)\phi)(\mathbf{x}) = \phi(g^t\mathbf{x}), \tag{1.156}$$

where $g^t$ is the phase flow, i.e. $g^t\mathbf{x}_0$ is the solution of (1.155).

The function defined as

$$\phi(\mathbf{x}, t) = (U(t)\phi)(\mathbf{x})$$

obey the following evolution equation:

$$\frac{\partial}{\partial t}\phi(\mathbf{x}, t) = X_{\mathbf{F}}(\mathbf{x})\phi(\mathbf{x}, t), \qquad \phi(\mathbf{x}, 0) = \phi(\mathbf{x}), \tag{1.157}$$

where $X_{\mathbf{F}}(\mathbf{x})$ is the vector field corresponding to (1.155)

$$X_{\mathbf{F}}(\mathbf{x}) = \mathbf{F}(\mathbf{x})\cdot\frac{\partial}{\partial \mathbf{x}}.$$

Observe that we can can write (1.157) in the form

$$\frac{\partial}{\partial t}\phi(\mathbf{x}, t) = \mathcal{M}^\dagger\phi(\mathbf{x}, t), \qquad \phi(\mathbf{x}, 0) = \phi(\mathbf{x}),$$

where the operator $\mathcal{M}^\dagger$ is given by

$$\mathcal{M}^\dagger = \mathbf{F}(\mathbf{a}^\dagger)\cdot\mathbf{a},$$

where $\mathbf{a}^\dagger$, $\mathbf{a}$ are the Bargmann realizations of the Bose operators (see appendix C):

$$(\mathbf{a}^\dagger\phi)(\mathbf{x}) = \mathbf{x}\phi(\mathbf{x}), \qquad (\mathbf{a}\phi)(\mathbf{x}) = \frac{\partial}{\partial \mathbf{x}}\phi(\mathbf{x}). \tag{1.158}$$

Proceeding analogously as in the case of eq. (1.153) one finds that the function $\phi(\mathbf{x}, t)$ satisfying (1.157) can be expressed by

$$\phi(\mathbf{x}, t) = \langle \mathbf{x}, t|\phi\rangle e^{\frac{1}{2}\mathbf{x}^2}, \tag{1.159}$$

where the vectors $|\mathbf{x}, t\rangle$ are the solution of the following abstract equation in Hilbert space:

$$\frac{d}{dt}|x,t\rangle = M|x,t\rangle, \qquad |x,0\rangle = |\mathbf{x}\rangle, \tag{1.160}$$

where $M = \mathbf{a}^{\dagger}\cdot\mathbf{F}(\mathbf{a})$ and $|\mathbf{x}\rangle$ is a normalized coherent state.

Evidently, (1.160) coincides with the Schrödinger-like equation corresponding to the system (1.155) which was introduced in section 1.1. Thus the actual treatment and the linearization based on (1.156) are related by (1.159).

REMARK. Note that (1.157) is nothing but the Liouville equation corresponding to (1.155). The solution of the system (1.155) is linked to the solution of (1.157) by

$$(g^t\mathbf{x})_i = \phi_i(\mathbf{x}, t), \qquad i = 1, \ldots, k, \tag{1.161}$$

where $\phi_i(\mathbf{x}, t)$ is the solution of (1.157) subject to the initial data $\phi_i(\mathbf{x}, 0) = x_i$. The Hilbert space generalization of (1.161) was discussed in section 1.4.1 (see (1.113)). ∎

CHAPTER 2

# PARTIAL DIFFERENTIAL EQUATIONS

## 2.1 Evolution equation in Hilbert space

In this chapter we generalize the Hilbert space approach developed in the previous chapter to the case of partial differential equations. The aim of this section is to introduce the linear, abstract evolution equation in Hilbert space which is a counterpart of systems of nonlinear partial differential equations of the evolution type with analytic nonlinearities. Consider the following system of nonlinear differential equations (complex or real):

$$\partial_t \boldsymbol{\xi}(x,t) = \mathbf{F}(\boldsymbol{\xi}, D^\alpha \boldsymbol{\xi}; x, t), \qquad \boldsymbol{\xi}(x,0) = \boldsymbol{\xi}_0(x), \tag{2.1}$$

where $\boldsymbol{\xi}$: $\mathbf{R}^s \times \mathbf{R} \to \mathbf{C}^k$, $D^\alpha\boldsymbol{\xi} = (D^{\alpha_1}\xi_1, \ldots, D^{\alpha_k}\xi_k)$, $\alpha_i$ are multiindices, $D^\beta = \partial^{|\beta|}/\partial x_1^{\beta_1}\cdots\partial x_s^{\beta_s}$, with $|\beta| = \sum_{i=1}^{s}\beta_i$, is a generalized derivative, $\mathbf{F}$ is analytic in $\boldsymbol{\xi}$, $D^\alpha\boldsymbol{\xi}$ and $\boldsymbol{\xi}_0 \in \bigoplus_{i=1}^{k} L^2(\mathbf{R}^s, d^s x)$, that is $\xi_{0i} \in L^2(\mathbf{R}^s, d^s x)$, $i = 1, \ldots, k$.

Let us assume that the solution $\boldsymbol{\xi}$ of the system (2.1) is square integrable. As with ordinary differential equations, we introduce the vectors of the form

$$|\xi, t\rangle = \exp\left[\frac{1}{2}\left(\int d^s x\, |\boldsymbol{\xi}|^2 - \int d^s x\, |\boldsymbol{\xi}_0|^2\right)\right] |\boldsymbol{\xi}\rangle \tag{2.2a}$$

$$= \exp\left(-\frac{1}{2}\int d^s x\, |\boldsymbol{\xi}_0|^2\right) \exp\left(\int d^s x\, \boldsymbol{\xi}(x,t)\cdot\mathbf{a}^\dagger(x)\right) |\mathbf{0}\rangle, \tag{2.2b}$$

where $|\boldsymbol{\xi}\rangle$ is a normalized functional coherent state (see appendix C), $\boldsymbol{\xi}$ fulfils (2.1), $a_i^\dagger(x)$, $x \in \mathbf{R}^s$, $i = 1, \ldots, k$, are the standard Bose field creation operators and $|\mathbf{0}\rangle$ is the vacuum vector.

By differentiating (2.2b) and taking into account (C.58) we find that the vectors (2.2) obey the following linear, Schrödinger like evolution equation in Hilbert space:

$$\frac{d}{dt}|\xi, t\rangle = M(t)|\xi, t\rangle, \qquad |\xi, 0\rangle = |\boldsymbol{\xi}_0\rangle, \tag{2.3}$$

with the boson "Hamiltonian" such that

$$M(t) = \int d^s x\, \mathbf{a}^\dagger(x)\cdot\mathbf{F}(\mathbf{a}(x), D^\alpha\mathbf{a}(x); x, t), \tag{2.4}$$

where $\mathbf{a}^\dagger(x)$, $\mathbf{a}(x)$, $x \in \mathbf{R}^s$, are the standard Bose field creation and annihilation operators, respectively.

Furthermore, using (2.2a) and (C.58) we obtain the following eigenvalue equation relating the solution $\boldsymbol{\xi}[\boldsymbol{\xi}_0|x,t]$ of the system (2.1) and the solution $|\boldsymbol{\xi}_0,t\rangle$ of (2.3):

$$\mathbf{a}(x)|\boldsymbol{\xi}_0,t\rangle = \boldsymbol{\xi}[\boldsymbol{\xi}_0|x,t]|\boldsymbol{\xi}_0,t\rangle. \tag{2.5}$$

We have thus shown that the integration of the nonlinear system (2.1), where $\boldsymbol{\xi}$ is square integrable, can be cast into the solution of the linear, Schrödinger-like evolution equation in Hilbert space (2.3). The postulate that the solutions to (2.1) are square integrable at any time, that is if $\boldsymbol{\xi}_0(x)$ is square integrable, then $\boldsymbol{\xi}[\boldsymbol{\xi}_0|x,t]$ is also square integrable, seems to be rather restrictive one. However, the class of equations satisfying this requirement is large enough to include such equations of classical and of current interest as Burgers equation, Korteweg-de Vries equation, nonlinear Schrödinger equation and Kadomtsev-Petviashvili equation (see remark below). It should also be noted that numerous soliton solutions are supposed to vanish at spatial infinity, that is they are assumed to satisfy the necessary condition for the square integrability. The restriction to the square integrable initial data does not seem to be the most important one. Later on, we shall demonstrate that the formalism works also in the case with analytic initial data. This observation is consistent with the celebrated Cauchy-Kowalevski theorem on the existence and the uniqueness of the solutions to systems of partial differential equations with analytic nonlinearities and analytic initial data.

REMARK. The square integrability of solutions to Korteweg-de Vries equation, nonlinear Schrödinger equation and Kadomtsev-Petviashvili equation at any time follows from the fact that the squared norm of solutions to these equations is conserved during the time evolution. In the case of the Burgers equation such that

$$\partial_t u = \nu \partial_x^2 u - u\partial_x u, \tag{2.6}$$

where u: $\mathbf{R} \times \mathbf{R} \to \mathbf{R}$, the square integrability of the solution at any time is implied by the following inequality:

$$\frac{d}{dt}\int dx\, u^2 = -2\nu \int dx\, (\partial_x u)^2 \leq 0,$$

which can be easily obtained from (2.6). ■

We now return to eq. (2.3). On taking into account (2.2a) and (C.62) we arrive at the following formula relating the solution of (2.1) and the solution of (2.3):

$$\xi_i[\boldsymbol{\xi}_0|x,t] = \langle x_{(i)}|\boldsymbol{\xi}_0,t\rangle \exp\left(\frac{1}{2}\int d^s x\, |\boldsymbol{\xi}_0|^2\right), \qquad i = 1,\ldots,k, \tag{2.7}$$

where $|x_{(i)}\rangle = a_i^\dagger(x)|0\rangle$ is a basis vector of the coordinate representation (see appendix C). Note that (2.7) is a counterpart of the expression (1.9).

REMARK. We note that whereas in the case of ordinary differential equations discussed in previous chapter, we deal with a formalism resembling quantum mechanics, the Hilbert space formulation for partial differential equations introduced above has the structure analogous to quantum field theory. In particular, the eigenvalue equation (2.5) indicates the "quantization" scheme of the form $\boldsymbol{\xi} \to \mathbf{a}$, where $\boldsymbol{\xi}$ satisfies (2.1). ■

We now discuss the realization of the abstract equation (2.3) in the coordinate representation (see appendix C) and we show that the resulting linearization of the system (2.1) can be treated as a generalization of the Carleman embedding to the case of partial differential equations. Consider the evolution equation (2.3). On writing (2.3) in the coordinate representation (see (C.51)) we arrive at the following system of linear hierarchy equations:

$$\partial_t \xi_{\mathbf{n}}(x_{11},\ldots,x_{1n_1},\ldots,x_{k1},\ldots,x_{kn_k};t)$$
$$= \sum_{\mathbf{m}\in \mathbf{Z}_+^k} \left(\prod_{i=1}^{k} \frac{1}{m_i!}\right) \int \prod_{r=1}^{k}\prod_{j=1}^{m_r} d^s x'_{rj}\, \mathcal{M}_{\mathbf{nm}}(t)\xi_{\mathbf{m}}(x'_{11},\ldots,x'_{1m_1},\ldots,x'_{k1},\ldots,x'_{km_k};t), \quad (2.8a)$$

$$\xi_{\mathbf{n}}(x_{11},\ldots,x_{1n_1},\ldots,x_{k1},\ldots,x_{kn_k};0) = \left(\prod_{i=1}^{k}\prod_{j=1}^{n_i} \xi_{0i}(x_{ij})\right) \exp\left(-\frac{1}{2}\int d^s x\, |\boldsymbol{\xi}_0|^2\right), \quad (2.8b)$$

where

$$\xi_{\mathbf{n}}(x_{11},\ldots,x_{1n_1},\ldots,x_{k1},\ldots,x_{kn_k};t) = \langle x_{11},\ldots,x_{1n_1},\ldots,x_{k1},\ldots,x_{kn_k}|\xi,t\rangle$$

and

$$\mathcal{M}_{\mathbf{nm}}(t) = \langle x_{11},\ldots,x_{1n_1},\ldots,x_{k1},\ldots,x_{kn_k}|M(t)|x'_{11},\ldots,x'_{1m_1},\ldots,x'_{k1},\ldots,x'_{km_k}\rangle.$$

Note that the function $\xi_{\mathbf{n}}$ is symmetric with respect to the spatial variables. It should also be noted that the kernel $\mathcal{M}_{\mathbf{nm}}$ of the integral operator from the right-hand side of (2.8a) is a distribution. Now, using (2.2a) and (C.62) we obtain the following relation:

$$\xi_{\mathbf{n}}(x_{11},\ldots,x_{1n_1},\ldots,x_{k1},\ldots,x_{kn_k};t) = \left(\prod_{i=1}^{k}\prod_{j=1}^{n_i} \xi_i(x_{ij},t)\right) \exp\left(-\frac{1}{2}\int d^s x\, |\boldsymbol{\xi}_0|^2\right). \quad (2.9)$$

Consequently, the solution of the system (2.1) is linked to the solution of (2.3) by (see also (2.7)):

$$\xi_i(x,t) = \xi_{\mathbf{e}_i}(x,t)\exp\left(\frac{1}{2}\int d^s x\, |\boldsymbol{\xi}_0|^2\right), \qquad i = 1,\ldots,k, \quad (2.10)$$

where $\mathbf{e}_i = (0,\ldots,0,1_i,0,\ldots,0)$ is the unit vector.

It thus appears that we in fact embedded the nonlinear system of partial differential equations (2.1) into the infinite system of linear hierarchy equations (2.8).

Comparing (1.20), (1.21) and (2.9), (2.10), respectively we find that such embedding is a very natural development of the approach taken by Carleman for linearization of ordinary differential equations.

Motivated by the importance of one-dimensional real systems (2.1) such that

$$\partial_t u(x,t) = F(u, D^\alpha u; x, t), \qquad u(x,0) = u_0(x), \tag{2.11}$$

where $u\colon \mathbf{R}^s \times \mathbf{R} \to \mathbf{R}$, $D^\alpha = \partial^{|\alpha|}/\partial x_1^{\alpha_1} \cdots \partial x_s^{\alpha_s}$, $|\alpha| = \sum_{i=1}^{s} \alpha_i$, $F$ is analytic in $u$, $D^\alpha u$ and $u_0 \in L^2_{\mathbf{R}}(\mathbf{R}^s, d^s x)$ (real Hilbert space of square integrable functions),

we now rewrite the above formulae in the case of (2.11). We have

$$\partial_t u_n(x_1, \ldots, x_n; t) = \sum_m \frac{1}{m!} \int d^s x_1' \ldots d^s x_m' \, \mathcal{M}_{nm}(t) u_m(x_1', \ldots, x_m'; t), \tag{2.12}$$

$$u_n(x_1, \ldots, x_n; 0) = \left( \prod_{i=1}^{n} u_0(x_i) \right) \exp\left( -\frac{1}{2} \int d^s x \, u_0^2 \right), \tag{2.13}$$

where

$$u_n(x_1, \ldots, x_n; t) = \langle x_1, \ldots, x_n | u, t \rangle \tag{2.14}$$

and

$$\mathcal{M}_{nm}(t) = \langle x_1, \ldots, x_n | M(t) | x_1', \ldots, x_m' \rangle. \tag{2.15}$$

The relations (2.9) and (2.10) reduce to

$$u_n(x_1, \ldots, x_n; t) = \left( \prod_{i=1}^{n} u(x_i, t) \right) \exp\left( -\frac{1}{2} \int d^s x \, u_0^2 \right), \tag{2.16}$$

$$u(x,t) = u_1(x,t) \exp\left( \frac{1}{2} \int d^s x \, u_0^2 \right). \tag{2.17}$$

REMARK. An alternative realization for the abstract equation (2.3) is the Bargmann one, when it takes the form of the linear equation in functional derivatives. As a matter of fact, the linear hierarchy equations and linear equations in functional derivatives obtained from nonlinear partial differential equations can be found in mathematical physics. Examples include Friedman-Keller hierarchy equations and the Hopf equation in functional derivatives [30] resulting from nonlinear Navier-Stokes equations. It must be realized, however, that such correspondences are only treated as the formal properties of particular nonlinear equations. ■

EXAMPLE. Consider the one-dimensional real equation

$$\partial_t u = -u \partial_x u, \qquad u(x,0) = u_0(x) \in L^2_{\mathbf{R}}(\mathbf{R}, dx). \tag{2.18}$$

It is easy to verify that the squared norm of the solution to (2.18) is conserved during the time evolution. Therefore, the Hilbert space approach described above can be applied to (2.18). The Schrödinger-like equation corresponding to (2.18) has the following form:

$$\frac{d}{dt}|u,t\rangle = M|u,t\rangle, \qquad |u,0\rangle = |u_0\rangle,$$

where the "Hamiltonian" (2.4) is

$$M = -\int dx\, a^{\dagger}(x)a(x)a'(x). \tag{2.19}$$

An easy calculation using (2.19), (C.57) and (C.55) shows that the kernel (2.15) is given by

$$\mathcal{M}_{nm} = \frac{1}{2}\delta_{n+1m}\left(\prod_{i=1}^{n}\delta(x_i' - x_i)\right)\sum_{i=1}^{n}\frac{\partial}{\partial x'_{n+1}}\delta(x'_{n+1} - x_i). \tag{2.20}$$

Inserting (2.20) into (2.12) yields

$$\partial_t u_n(x_1,\ldots,x_n;t) = -\frac{1}{2}\sum_{i=1}^{n}\partial_{x_i}u_{n+1}(x_1,\ldots,x_n,x_i;t). \tag{2.21}$$

On expanding the solution of (2.21) in the power series of $t$

$$u_n(x_1,\ldots,x_n;t) = \sum_{i=0}^{\infty} c_{ni}(x_1,\ldots,x_n)t^i \tag{2.22}$$

and setting $x_i = x$, $i = 1, \ldots, n$, we obtain from (2.21) the following differential-difference equation:

$$-\frac{n}{n+1}c'_{n+1i} = (i+1)c_{ni+1}. \tag{2.23}$$

It can be easily checked that the solution of (2.23) is

$$c_{ni} = \frac{(-1)^i}{i!}\frac{n}{n+i}c^{(i)}_{n+i0}. \tag{2.24}$$

On putting in (2.24) $n = 1$ and using (2.22) together with (2.17) we arrive at the following solution of (2.18):

$$u[u_0|x,t] = \sum_{i=0}^{\infty}\frac{(-t)^i}{(i+1)!}(u_0^{i+1})^{(i)}. \tag{2.25}$$

It is easy to verify that whenever $x$, $t$ are such that $\partial_t u$ is nonsingular (this requirement leads to the condition $tu_0'(\sigma)|_{\sigma=x-tu} \neq -1$), then (2.25) is the Maclaurin series for the implicit solution to (2.18) of the form

$$u - u_0(x - tu) = 0. \tag{2.26}$$

It should be noted that the formulae (2.25) and (2.26) hold true regardless of the square integrability of $u_0$. For example, the function such that

$$u[x|x,t] = \frac{x}{1+t}$$

obtained from (2.25) or (2.26), where $u_0 = x$, is easily shown to be the solution of (2.18). □

## 2.2 Operator evolution equations

As we have already demonstrated in previous section, an advantage of the actual "quantal" approach is that it permits application of constructions described in chapter 1 to the study of nonlinear partial differential equations. Our main concern in this section is to extend the observations of section 1.2 to the case of the systems (2.1). Consider the "Heisenberg equations of motion" satisfied by the time-dependent Bose annihilation operators

$$\partial_t \mathbf{a}(x,t) = [\mathbf{a}(x,t), M_{\mathrm{H}}(t)], \qquad \mathbf{a}(x,0) = \mathbf{a}(x), \tag{2.27}$$

where $M_{\mathrm{H}}(t) = V(t)^{-1}M(t)V(t)$ is the "Hamiltonian" in "Heisenberg picture". Here the evolution operator $V(t)$ is the solution of the operator evolution equation

$$\frac{dV}{dt} = M(t)V, \qquad V(0) = I, \tag{2.28}$$

where $M(t)$ is given by (2.4).

It is clear that the solution of (2.28) is linked to the solution of (2.3) by

$$|\boldsymbol{\xi}_0, t\rangle = V(t)|\boldsymbol{\xi}_0\rangle.$$

The formal solution to (2.27) is of the form

$$\mathbf{a}(x,t) = V(t)^{-1}\mathbf{a}(x)V(t).$$

Now, taking into account the relation (see (1.34)):

$$\mathbf{a}(x,t)|\boldsymbol{\xi}_0\rangle = \boldsymbol{\xi}[\boldsymbol{\xi}_0|x,t]|\boldsymbol{\xi}_0\rangle$$

and proceeding analogously as with ordinary differential equations we arrive at the following formal solution of the system (2.1):

$$\begin{aligned} \boldsymbol{\xi}[\boldsymbol{\xi}_0|x,t] &= \langle\boldsymbol{\xi}_0|\mathbf{a}(x,t)|\boldsymbol{\xi}_0\rangle \\ &= \boldsymbol{\xi}_0(x) + \sum_{i=1}^{\infty}\frac{(-1)^i}{i!}\langle\boldsymbol{\xi}_0|[\Phi(t),\ldots,[\Phi(t),\mathbf{a}(x)]\ldots]|\boldsymbol{\xi}_0\rangle, \end{aligned} \tag{2.29}$$

where $\Phi(t)$ is the "phase operator" given by (1.31) and (2.4).

In particular case of the autonomous system (2.1) such that

$$\partial_t\boldsymbol{\xi}(x,t) = \mathbf{F}(\boldsymbol{\xi}, D^\alpha\boldsymbol{\xi}; x), \qquad \boldsymbol{\xi}(x,0) = \boldsymbol{\xi}_0(x), \tag{2.30}$$

where $\boldsymbol{\xi}$: $\mathbf{R}^s \times \mathbf{R} \to \mathbf{C}^k$, $\mathbf{F}$ is analytic in $\boldsymbol{\xi}$, $D^\alpha\boldsymbol{\xi}$ and $\boldsymbol{\xi}_0 \in \bigoplus_{i=1}^k L^2(\mathbf{R}^s, d^sx)$, the "phase operator" is $\Phi(t) = tM$, where

$$M = \int d^sx\, \mathbf{a}^\dagger(x)\cdot\mathbf{F}(\mathbf{a}(x), D^\alpha\mathbf{a}(x); x)$$

is the "Hamiltonian" corresponding to (2.30).

Therefore, the solution of (2.30) given by (2.29) takes the form of the formal power series in $t$

$$\boldsymbol{\xi}[\boldsymbol{\xi}_0|x,t] = \boldsymbol{\xi}_0(x) + \sum_{i=1}^{\infty} \frac{(-t)^i}{i!} \langle \boldsymbol{\xi}_0 | [M, \ldots, [M, \mathbf{a}(x)] \ldots ] | \boldsymbol{\xi}_0 \rangle. \tag{2.31}$$

An easy inspection based on (C.46a) shows that (2.31) coincides with the formal solution to (2.30) given by the Lie series of the form

$$\boldsymbol{\xi}[\boldsymbol{\xi}_0|x,t] = \sum_{i=0}^{\infty} \frac{t^i}{i!} \left[ \mathbf{F}(\boldsymbol{\xi}_0, D^\alpha \boldsymbol{\xi}_0; x) \cdot \frac{\partial}{\partial \boldsymbol{\xi}_0} + \sum_\alpha D^\alpha \mathbf{F}(\boldsymbol{\xi}_0, D^\alpha \boldsymbol{\xi}_0; x) \cdot \frac{\partial}{\partial D^\alpha \boldsymbol{\xi}_0} \right]^i \boldsymbol{\xi}_0(x). \tag{2.32}$$

It thus appears, that, as with ordinary differential equations, the Lie series approach to the study of (2.30) corresponds to the "Heisenberg picture" within the actual "quantal" Hilbert space formalism.

EXAMPLE. Consider equation (2.18). An easy computation using (C.53) gives the relation

$$\underbrace{[M, \ldots, [}_{i-\text{times}} M, a(x)] \ldots ] = \frac{1}{i+1} [a(x)^{i+1}]^{(i)}.$$

Hence, taking into account (2.31) we obtain the solution (2.25) of (2.18). □

Consider now the real autonomous system (2.1)

$$\partial_t \mathbf{u}(x,t) = \mathbf{F}(\mathbf{u}, D^\alpha \mathbf{u}; x), \qquad \mathbf{u}(x,0) = \mathbf{u}_0(x), \tag{2.33}$$

where $\mathbf{u}\colon \mathbf{R}^s \times \mathbf{R} \to \mathbf{R}^k$, $\mathbf{F}$ is analytic in $\mathbf{u}$, $D^\alpha \mathbf{u}$ and $\mathbf{u}_0 \in \bigoplus_{i=1}^k L^2_{\mathbf{R}}(\mathbf{R}^s, d^s x)$.

Using the algorithm described in section 1.2 (see (1.40)) we find that (2.33) can be cast into the following linear recurrence in Hilbert space:

$$|i, r+1\rangle = M^\dagger |i, r\rangle, \qquad |i, 0\rangle = |x_{(i)}\rangle, \tag{2.34}$$

where $M^\dagger$ is the Hermitian conjugate of the "Hamiltonian" corresponding to (2.33), that is,

$$M^\dagger = \int d^s x\, \mathbf{F}(\mathbf{a}^\dagger(x), D^\alpha \mathbf{a}^\dagger(x); x) \cdot \mathbf{a}(x)$$

and $|x_{(i)}\rangle = a_i^\dagger(x)|\mathbf{0}\rangle$ is a basis vector of the coordinate representation (see (2.7)).

The solution of (2.33) and the solution of (2.34) are related by

$$u_i[\mathbf{u}_0|x,t] = \sum_{r=0}^{\infty} \frac{t^r}{r!} \langle i, r | \mathbf{u}_0 \rangle \exp\left( \frac{1}{2} \int d^s x\, \mathbf{u}_0^2 \right), \qquad i = 1, \ldots, k. \tag{2.35}$$

Writing (2.34) in the Bargmann representation (see (C.64) and (C.67)) and using (2.35) we obtain the solution to (2.33) of the form

$$\mathbf{u}[\mathbf{u}_0|x,t] = \sum_{r=0}^{\infty} \frac{t^r}{r!} \left( \int d^s x'\, \mathbf{F}(\mathbf{u}_0(x'), D^\alpha \mathbf{u}_0(x'); x') \cdot \frac{\delta}{\delta \mathbf{u}_0(x')} \right)^r \mathbf{u}_0(x), \tag{2.36}$$

where $\delta/\delta \mathbf{u}_0$ designates the functional derivative (see appendix E).

Taking into account (E.12) we find that (2.36) coincides with the Lie series expansion for (2.33) given by (2.32). Thus, it turns out that the standard theory of Lie series based on the formulae like (2.32) corresponds to the Bargmann representation within the introduced canonical Hilbert space approach. On the other hand, in view of (1.46), the series (2.36) seems to be the most natural generalization of a Lie series in the case of nonlinear partial differential equations.

## 2.3 Symmetries and first integrals

We now use the analysis given in section 1.3 to develop symmetries and first integrals for partial differential equations within the actual Hilbert space formalism [31]. We examine first the symmetries. Consider the real, autonomous, analytic system of the form

$$\partial_t \mathbf{u}(x,t) = \mathbf{F}(\mathbf{u}, D^\alpha \mathbf{u}), \tag{2.37}$$

where $\mathbf{u}$: $\mathbf{R}^s \times \mathbf{R} \to \mathbf{R}^k$, $\mathbf{F}$ is analytic in $\mathbf{u}$, $D^\alpha \mathbf{u}$ and $\mathbf{u}_0 \in \bigoplus_{i=1}^k L^2_{\mathbf{R}}(\mathbf{R}^s, d^s x)$.

Recall that the function $\boldsymbol{\sigma}(\mathbf{u}) \equiv \boldsymbol{\sigma}(\mathbf{u}, D^\beta \mathbf{u})$ is a symmetry of (2.37) if it leaves (2.37) invariant within order $\epsilon$, i.e. the equation

$$\partial_t \mathbf{u}'(x,t) = \mathbf{F}(\mathbf{u}'),$$

where $\mathbf{F}(\mathbf{u}) \equiv \mathbf{F}(\mathbf{u}, D^\alpha \mathbf{u})$,

must be correct to order $\epsilon$. Taking into account the following relations:

$$\partial_t \boldsymbol{\sigma} = \boldsymbol{\sigma}'[\mathbf{F}], \qquad \mathbf{F}(\mathbf{u} + \epsilon\boldsymbol{\sigma}) = \mathbf{F}(\mathbf{u}) + \epsilon \mathbf{F}'[\boldsymbol{\sigma}] + \mathcal{O}(\epsilon^2),$$

where prime designates the Gateaux derivative (see appendix E), we find that the symmetry satisfies

$$\mathbf{F}'[\boldsymbol{\sigma}] = \boldsymbol{\sigma}'[\mathbf{F}]. \tag{2.38}$$

Using the Lie bracket of analytic vector fields $\mathbf{f}$, $\mathbf{g}$ on the space of vector-valued functions on $\mathbf{R}^s$ vanishing sufficiently fast at the spatial infinity, defined by

$$[\mathbf{f}, \mathbf{g}] = \mathbf{f}'[\mathbf{g}] - \mathbf{g}'[\mathbf{f}], \tag{2.39}$$

we arrive at the following form of the condition (2.38):

$$[\mathbf{F}, \boldsymbol{\sigma}] = 0. \tag{2.40}$$

We now return to the Hilbert space approach. Let us introduce the following mapping:

$$\mathbf{f} \to L_{\mathbf{f}} = \int d^s x\, \mathbf{a}^\dagger(x) \cdot \mathbf{f}(\mathbf{a}(x)), \tag{2.41}$$

where $\mathbf{f}(\mathbf{a}) \equiv \mathbf{f}(\mathbf{a}, D^\gamma \mathbf{a})$.

Taking into account (C.46a) and (E.11) we see that (2.41) defines an isomorphism between Lie algebra of vector fields with the bracket (2.39) and Lie algebra of boson field operators linear in creation operators, i.e.,

$$[L_{\mathbf{f}}, L_{\mathbf{g}}] = L_{[\mathbf{f},\mathbf{g}]}. \tag{2.42}$$

Using (2.41) and (2.42) we obtain the following form of the condition (2.40) for $\boldsymbol{\sigma}(\mathbf{u})$ to be a symmetry of the system (2.37):

$$[M, \Sigma] = 0, \tag{2.43}$$

where $M = \int d^s x\, \mathbf{a}^\dagger(x)\cdot\mathbf{F}(\mathbf{a}(x))$ is the "Hamiltonian" corresponding to the system (2.37), and

$$\Sigma = \int d^s x\, \mathbf{a}^\dagger(x)\cdot\boldsymbol{\sigma}(\mathbf{a}(x)). \tag{2.44}$$

It thus appears that the symmetries of the system (2.37) correspond within the actual "quantal" Hilbert space formalism to operators of the form (2.44) which commute with the "Hamiltonian" $M$.

Let us recall that in the case of ordinary differential equations the standard theory based on the concept of a vector field corresponds to the particular Bargmann representation within the canonical Hilbert space approach. We now demonstrate that the same holds true in the case of the actual treatment concerning partial differential equations. Consider the operators $X_{\mathbf{f}}$ defined by

$$X_{\mathbf{f}}\mathbf{g} = \mathbf{g}'[\mathbf{f}].$$

Taking into account the following relation (see (E.14)):

$$X_{\mathbf{f}} = \int d^s x\, \mathbf{f}(\mathbf{u})\cdot\frac{\delta}{\delta \mathbf{u}}, \tag{2.45}$$

where $\delta/\delta\mathbf{u}$ designates the functional derivative (see appendix E),

we see that the operators $X_{\mathbf{f}}$ are the natural generalizations of the vector fields corresponding to ordinary differential equations (see (1.73)). In particular, the counterpart of the relation (1.74) is

$$[X_{\mathbf{f}}, X_{\mathbf{g}}] = X_{\mathbf{g}'[\mathbf{f}]-\mathbf{f}'[\mathbf{g}]}. \tag{2.46}$$

The derivation of (2.46) is straightforward, using (E.11), (E.13) and the fact that the functional derivatives satisfy the Leibnitz's rule. It follows immediately from (2.46) that the condition (2.38) for $\boldsymbol{\sigma}(\mathbf{u})$ to be a symmetry of the system (2.37) can be written as

$$[X_{\mathbf{F}}, X_{\boldsymbol{\sigma}}] = 0, \tag{2.47}$$

where $X_{\mathbf{F}} = \int d^s x\, \mathbf{F}(\mathbf{u})\cdot\frac{\delta}{\delta\mathbf{u}}$ and $X_{\boldsymbol{\sigma}} = \int d^s x\, \boldsymbol{\sigma}(\mathbf{u})\cdot\frac{\delta}{\delta\mathbf{u}}$.

We now return to the Hilbert space formalism. Consider the following mapping:

$$X_{\mathbf{f}} \to L_{\mathbf{f}}^{\dagger} = \int d^s x\, \mathbf{f}(\mathbf{a}^{\dagger}(x))\cdot\mathbf{a}(x). \tag{2.48}$$

Using (C.46b), (E.11) and (2.45) we find that under the map (2.48) the commutator (2.46) transforms as

$$[X_{\mathbf{f}}, X_{\mathbf{g}}] \to [L_{\mathbf{f}}^{\dagger}, L_{\mathbf{g}}^{\dagger}]. \tag{2.49}$$

Therefore, the mapping (2.48) establishes isomorphism of the Lie algebra of operators with the bracket (2.46) and Lie algebra of boson field operators linear in annihilation operators. Equations (2.47) and (2.49) taken together yield the following form of the symmetry condition (2.38):

$$[M^{\dagger}, \Sigma^{\dagger}] = 0, \tag{2.50}$$

where $M^{\dagger} = \int d^s x\, \mathbf{F}(\mathbf{a}^{\dagger})\cdot\mathbf{a}$ and $\Sigma^{\dagger} = \int d^s x\, \boldsymbol{\sigma}(\mathbf{a}^{\dagger})\cdot\mathbf{a}$,

that is the conjugate of (2.43) is obtained. It is clear that (2.50) written in the Bargmann representation (see appendix C) coincides with (2.47). In fact, the Bose field operators act in this representation as follows (see formulae (C.67)):

$$\mathbf{a}^{\dagger} = \mathbf{u}, \qquad \mathbf{a} = \frac{\delta}{\delta \mathbf{u}}.$$

We have thus shown that the standard approach to the theory of symmetries of partial differential equations based on the concept of a vector field and the Gateaux derivative is included by the canonical Hilbert space formalism as a particular case of the Bargmann representation. The above observations concerning symmetries of partial differential equations are applied in section 4.1, where the Hilbert space counterpart of a master symmetry is investigated.

We now discuss the first integrals (conservation laws). Consider the system (2.37). Recall that the functional $I[\mathbf{u}]$ is said to be the first integral for (2.37) if it satisfies

$$\int d^s x\, \mathbf{F}(\mathbf{u})\cdot\frac{\delta I[\mathbf{u}]}{\delta \mathbf{u}} = I'[\mathbf{F}] = 0. \tag{2.51}$$

Using (E.3) we find that the total time derivative of $I[\mathbf{u}]$ along the solution of (2.37) vanishes, that is,

$$\frac{dI[\mathbf{u}]}{dt} = 0.$$

Now let $M$ be the "Hamiltonian" corresponding to the system (2.37) such that

$$M = \int d^s x\, \mathbf{a}^{\dagger}(x)\cdot\mathbf{F}(\mathbf{a}(x)). \tag{2.52}$$

Assuming that $I[\mathbf{u}]$ is analytic in $\mathbf{u}$ and taking into account (2.52) and (C.47) we obtain the following equivalent form of (2.51):

$$[M, I[\mathbf{a}]] = 0. \tag{2.53}$$

Thus, it turns out that the first integrals for the system (2.37) are represented by operator invariants which are functionals depending solely on Bose annihilation operators.

We now derive the vector counterpart of the relation (2.53). Let $I[\mathbf{u}]$ be the first integral of (2.37). Proceeding analogously as in the case of ordinary differential equations we introduce the vectors of the form

$$|\phi\rangle = I[\mathbf{a}^\dagger]|\mathbf{0}\rangle. \tag{2.54}$$

It is easy to verify that the vectors (2.54) fulfil the following equation in Hilbert space:

$$M^\dagger|\phi\rangle = 0, \tag{2.55}$$

where $M^\dagger$ is the Hermitian conjugate of the "Hamiltonian" (2.52).

Projecting (2.55) onto the normalized functional coherent state $|\mathbf{u}\rangle$ and using (C.64) and (C.65) we find that the first integrals for (2.37) are linked to the solution of (2.55) by

$$I[\mathbf{u}] = \langle\mathbf{u}|\phi\rangle \exp\left(\frac{1}{2}\int d^s x\, \mathbf{u}^2\right)$$

It thus appears that the problem of determining analytic first integrals of (2.37) can be reduced to the solution of the abstract equation (2.55). We note that the original definition (2.51) corresponds to the particular Bargmann representation for the abstract equation (2.55). An application of (2.55) can be found in section 4.1, where the method of determining the first integrals is described.

CHAPTER 3

# DIFFERENCE EQUATIONS

## 3.1 Evolution equation in Hilbert space

Our concern so far has been exclusively with continuous-time dynamical systems. The goal of this chapter is to introduce a generalization of the Hilbert space approach to the case of nonlinear difference equations. We begin by deriving the linear, Hilbert space counterpart of the following nonlinear recurrence:

$$x_{n+1} = f(x_n), \tag{3.1}$$

where $n \in \mathbf{Z}_+$ and $f\colon \mathbf{R} \to \mathbf{R}$ is analytic in $x_n$.

Following the scheme of the Hilbert space formalism developed in previous chapters we define the vectors such that

$$\begin{aligned} |x,n\rangle &= \exp[\tfrac{1}{2}(x_n^2 - x_0^2)]|x_n\rangle & (3.2a)\\ &= \exp(-\tfrac{1}{2}x_0^2)\exp(x_n a^\dagger)|0\rangle, & (3.2b) \end{aligned}$$

where $x_n$ fulfils (3.1) and $|x_n\rangle$ is a normalized coherent state (see appendix C).

Taking into account (3.1) and (3.2b) we obtain the following linear difference equation in Hilbert space:

$$|x,n+1\rangle = M|x,n\rangle, \qquad |x,0\rangle = |x_0\rangle, \tag{3.3}$$

where $M$ is a boson operator of the form

$$M = \sum_{i=0}^{\infty} \frac{1}{i!} a^{\dagger i}[f(a) - a]^i. \tag{3.4}$$

Here $a^\dagger$, $a$ are the standard Bose creation and annihilation operators. Furthermore, it follows immediately from (3.2a) that the following eigenvalue equation holds true:

$$a|x_0,n\rangle = x_n(x_0)|x_0,n\rangle, \tag{3.5}$$

where $|x_0,n\rangle$ designates the solution of (3.3) and $x_n(x_0)$ is the solution of (3.1).

Thus, it turns out that the solution of the nonlinear recurrence (3.1) can be reduced to the solution of the linear, abstract difference equation in Hilbert space (3.3).

REMARK. It should be mentioned that the approach can be immediately generalized to include nonautonomous complex systems of nonlinear difference equations. Motivated by the importance of some two-dimensional autonomous systems such as, for example, the Hénon one, we now write down the linear recurrence in Hilbert space corresponding to the system

$$\mathbf{x}_{n+1} = \mathbf{F}(\mathbf{x}_n), \tag{3.6}$$

where $n \in \mathbf{Z}_+$ and $\mathbf{F}\colon \mathbf{R}^2 \to \mathbf{R}^2$. Following the scheme presented above one finds easily that (3.6) can be brought down to the linear recurrence in Hilbert space such that

$$|x, n+1\rangle = M|x, n\rangle,$$

where the boson displacement operator $M$ is

$$M = \sum_{i=0}^{\infty} \frac{1}{i!} \sum_{j=0}^{i} \binom{i}{j} a_1^{\dagger i-j} a_2^{\dagger j} (F_1(\mathbf{a}) - a_1)^{i-j} (F_2(\mathbf{a}) - a_2)^j. \qquad ■$$

We now return to (3.3). The reader may notice that the formal solution of (3.3) is given by

$$|x_0, n\rangle = M^n |x_0\rangle. \tag{3.7}$$

We also write down the following relation which is an immediate consequence of (3.2a) and (C.25):

$$x_n(x_0) = \langle 1|x_0, n\rangle \exp(\tfrac{1}{2}x_0^2), \tag{3.8}$$

where $x_n(x_0)$ is the solution of (3.1), $|x_0, n\rangle$ is the solution to (3.3) and $|1\rangle = a^\dagger|0\rangle$ is a basis vector of the occupation number representation (see appendix C).

EXAMPLE. Consider the logistic equation

$$x_{n+1} = \mu x_n(1 - x_n). \tag{3.9}$$

On using (3.4), (C.14) and (C.15) we arrive at the following form of the operator $M$ corresponding to (3.9):

$$M = \mu^N \sum_{i=0}^{\infty} \binom{N}{i} (-a)^i,$$

where $N = a^\dagger a$ is the number operator. It follows that (see (C.13)):

$$M^n = \mu^{nN} \sum_{i_1,\ldots,i_n} \binom{N}{i_1} \left[\prod_{r=1}^{n-1} \binom{N + \sum_{p=1}^{r} i_p}{i_{r+1}}\right] \mu^{\sum_{r=1}^{n-1}(n-r)i_r} (-a)^{\sum_{s=1}^{n} i_s}, \qquad n \geq 2. \tag{3.10}$$

Taking into account (3.7), (3.8) and (3.10) we obtain the following formal solution $x_n(x_0)$ to (3.9):

$$x_1(x_0) = \langle 1|M|x_0\rangle \exp(\tfrac{1}{2}x_0^2) = \mu x_0(1 - x_0),$$

$$x_n(x_0) = \langle 1|M^n|x_0\rangle \exp(\tfrac{1}{2}x_0^2)$$

$$= \mu^n x_0 \sum_{i_1,\ldots,i_n} \binom{1}{i_1} \left[\prod_{r=1}^{n-1} \binom{1 + \sum_{p=1}^{r} i_p}{i_{r+1}}\right] \mu^{\sum_{r=1}^{n-1}(n-r)i_r} (-x_0)^{\sum_{s=1}^{n} i_s}, \qquad n \geq 2. \qquad \square$$

As shown by Steeb [32], the Carleman embedding can be extended to nonlinear recurrences. We now demonstrate that the Carleman linearization approach to the nonlinear difference equations amounts a particular case of the actual tratment. Consider eq. (3.1). This equation written in the occupation number representation takes the following form:

$$x_{mn+1} = \sum_{m' \in \mathbf{Z}_+} M_{mm'} x_{m'n}, \tag{3.11a}$$

$$x_{m0} = \frac{x_0^m}{\sqrt{m!}} \exp(-\tfrac{1}{2}x_0^2), \tag{3.11b}$$

where $x_{mn} = \langle m|x,n\rangle$ and $M_{mm'} = \langle m|M|m'\rangle$.

Furthermore, taking into account (3.2a) and (C.25) we arrive at the following relation:

$$x_{mn} = \frac{x_n^m}{\sqrt{m!}} \exp(-\tfrac{1}{2}x_0^2). \tag{3.12}$$

Therefore, we have (see also (3.8)):

$$x_n = x_{1n} \exp(\tfrac{1}{2}x_0^2),$$

where $x_n$ satisfies (3.1).

We conclude that the finite-dimensional nonlinear recurrence (3.1) is embedded into the infinite-dimensional linear system of difference equations (3.11). Evidently, (3.12) is a version of the Carleman linearization ansatz (see (1.20) and (2.9)). It thus appears that the Carleman embedding approach to nonlinear recurrences is included by the canonical Hilbert space formalism introduced above as a particular case of the occupation number representation.

## 3.2 Operator evolution equations

As we have seen, the "Heisenberg picture" within the "quantal" approach to continuous-time dynamical systems corresponds to the technique of Lie series. Our aim in this section is to extend this observation to the case of difference equations. Consider the following operator recurrence:

$$V_{n+1} = MV_n, \qquad V_0 = I, \tag{3.13}$$

where $M$ is given by (3.4).

The obvious solution of (3.13) is

$$V_n = M^n. \tag{3.14}$$

Evidently, the operator equation (3.13) is equivalent to (3.3). Referring back to the derivation of (3.7), we recall that the solution $|x_0, n\rangle$ of (3.3) and the solution to (3.13) are related by

$$|x_0, n\rangle = V_n|x_0\rangle. \tag{3.15}$$

It is clear that $V_n$ plays the role of the evolution operator corresponding to the "Hamiltonian" $M$. We are now ready to derive the discrete-time counterpart of the "Heisenberg equations of motion". Consider equation (3.5). Equations (3.5), (3.3), (3.1) and (C.17) taken together imply

$$aM = Mf(a). \tag{3.16}$$

It might be observed that (3.16) corresponds to the "Heisenberg equations of motion" (1.130a). Prompted by this analogy, we write the "Heisenberg equations of motion" satisfied by the time-dependent Bose operators $a_n$ in the form

$$a_{n+1} = \langle a_n, M\rangle, \qquad a_0 = a, \tag{3.17}$$

where the operation $\langle a_n, \cdot\rangle$ is defined by

$$a_{n+1} = \langle a_n, B\rangle \quad \text{if and only if} \quad a_n B = B a_{n+1}.$$

Using (3.17) and (3.14) we find that the solution of (3.17) is linked to the solution of (3.13) by

$$aV_n = V_n a_n.$$

Hence, making use of (3.15) and (3.5) we obtain the following eigenvalue equation:

$$a_n|x_0\rangle = x_n(x_0)|x_0\rangle, \tag{3.18}$$

where $x_n(x_0)$ is the solution of (3.1).

On taking into account (3.18) and (3.17) we arrive at the following form of the solution to (3.1):

$$\begin{aligned} x_n(x_0) &= \langle x_0|a_n|x_0\rangle \\ &= \langle x_0|\underbrace{\langle \ldots \langle}_{n-\text{times}} a, M\rangle, \ldots, M\rangle|x_0\rangle. \end{aligned} \tag{3.19}$$

The reader may notice that (3.19) is a discrete-time counterpart of the Lie series discussed in previous chapters (compare (1.39) or (2.31)). Furthermore, using (3.17) and the following identity which is an immediate consequence of (3.16):

$$\phi(a)M = M\phi(f(a)), \tag{3.20}$$

where $\phi$ is an arbitrary analytic function,

we find that (3.19) coincides with the classical solution to (3.1) given by the functional iteration:

$$x_n(x_0) = \underbrace{f(f \ldots f(f}_{n-\text{times}}(x_0)) \ldots)). \tag{3.21}$$

We have thus shown that the classical formula (3.21) corresponds to the "Heisenberg picture" within the introduced "quantal" Hilbert space approach. On the other hand, it turns out that (3.21) is the most natural counterpart of a Lie series in the theory of difference equations.

We now return to (3.1). Proceeding analogously as in the case of continuous-time dynamical systems (see (1.40) and (2.33)) we find that (3.1) can be brought down to the linear difference equation in Hilbert space such that

$$|n+1\rangle_{\bullet} = M^{\dagger}|n\rangle_{\bullet}, \qquad |0\rangle_{\bullet} = |1\rangle, \tag{3.22}$$

where $M^{\dagger}$ is the Hermitian conjugate of the operator $M$ given by (3.4), i.e.,

$$M^{\dagger} = \sum_{i=0}^{\infty} \frac{1}{i!}(f(a^{\dagger}) - a^{\dagger})^{i} a^{i}, \tag{3.23}$$

and $|1\rangle = a^{\dagger}|0\rangle$ is a basis vector of the occupation number representation (see (3.8)).

Clearly, the solution to (3.1) can be recovered from the solution to (3.22) by

$$x_n(x_0) = {}_{\bullet}\langle n|x_0\rangle \exp(\tfrac{1}{2}x_0^2). \tag{3.24}$$

Now, (3.22) written in the Bargmann representation (see formulae (C.31) and (C.34)) takes the form of the linear functional equation

$$\tilde{\phi}_{n+1}(x_0) = \tilde{\phi}_n(f(x_0)), \qquad \tilde{\phi}_0(x_0) = x_0, \tag{3.25}$$

where $\tilde{\phi}_n(x_0) = \langle x_0|n\rangle_{\bullet} \exp(\frac{1}{2}x_0^2)$.

On using the solution of the trivial equation (3.25) and (3.24) we arrive at the solution (3.21). It thus appears that the classical formula (3.21) corresponds to the particular Bargmann representation in the introduced canonical Hilbert space formalism.

## 3.3 Functional equations

### *3.3.1 Abstract equation in Hilbert space*

This section deals with the Hilbert space description of the functional equations. As an illustration of the introduced formalism, we discuss the linearization transformations for nonlinear difference equations. Consider the following general functional equation:

$$\phi(f(x)) = \psi(x), \tag{3.26}$$

where the real-valued functions $\phi$, $f$, $\psi$ are analytic in $x \in \mathbf{R}$. By virtue of (3.20) this equation is equivalent to the following operator equation in Hilbert space:

$$\phi(a)M = M\psi(a) \tag{3.27}$$

where $M$ is given by (3.4).

We now define the vectors $|\phi\rangle$ and $|\psi\rangle$ as follows (see (C.32)):

$$|\phi\rangle = \phi(a^{\dagger})|0\rangle, \qquad |\psi\rangle = \psi(a^{\dagger})|0\rangle. \tag{3.28}$$

Taking the Hermitian conjugate of (3.27) and using (3.28) we find that these vectors satisfy the following equation:

$$M^\dagger|\phi\rangle = |\psi\rangle, \tag{3.29}$$

where $M^\dagger$ is the Hermitian conjugate of the operator $M$ given by (3.23).

The solution of (3.26) is related to the solution of (3.29) by

$$\phi(x) = \langle x|\phi\rangle \exp(\tfrac{1}{2}x^2), \qquad \psi(x) = \langle x|\psi\rangle \exp(\tfrac{1}{2}x^2),$$

where $|x\rangle$ is a normalized coherent state.

Thus, it turns out that the functional equation (3.26) can be reduced to the Hilbert space equation (3.29). On writing (3.29) in the occupation number representation and using the identity

$$Ma^\dagger = a^\dagger M f'(a)$$

following from (3.4) and (C.2a), we obtain

$$\phi(f_0) = \psi_0, \tag{3.30a}$$

$$\sum_{i=1}^{k} P_{ik}[f_0', \dots, f_0^{(k-i+1)}]\phi^{(i)}(f_0) = \psi_0^{(k)}, \qquad k = 1,\, 2,\, \dots,\, \infty, \tag{3.30b}$$

where $f_0^{(r)} = f^{(r)}(0)$, $\psi_0^{(k)} = \psi^{(k)}(0)$ and the polynomials $P_{ik}[f', \dots, f^{(k-i+1)}]$ are determined by the following recursion relations:

$$P_{1k} = f^{(k)}, \tag{3.31a}$$

$$P_{ik+1} = f' P_{i-1k} + P_{ik}', \qquad 1 < i < k+1, \tag{3.31b}$$

$$P_{kk} = f'^k \tag{3.31c}$$

We leave it to the reader to verify that the formulae (3.30) and (3.31) can be derived by taking the $n$th derivative of both sides of (3.26). It thus appears that the method of the power series (Maclaurin series) expansion for the solution of (3.26) is included by the introduced canonical Hilbert space formalism as a particular case of the occupation number representation for the abstract equation (3.29). It is worthwhile to point out that the recursive setting (3.30) and (3.31) for obtaining coefficients of a composite function is much more simple than the method of the substitution of a series into series or the Faa di Bruno formula. Finally, we note that (3.26) corresponds to the particular Bargmann representation for the abstract equation (3.29). Indeed, one finds easily that (3.29) written in the Bargmann representation coincides with (3.26).

### *3.3.2 Linearization transformations*

Our purpose now is to study the linearization transformations. We begin by recalling that the linearization transformation $\phi$ for eq. (3.1) satisfies the following functional equation:

$$\phi(f(x)) = \lambda\phi(x), \tag{3.32}$$

where $\phi$: $\mathbf{R} \to \mathbf{R}$ and $\lambda \neq 1$ is a constant.

In fact, using (3.1) we find that whenever (3.32) holds, then the transformation

$$x'_n = \phi(x_n) \tag{3.33}$$

converts (3.1) into the linear equation

$$x'_{n+1} = \lambda x'_n. \tag{3.34}$$

Suppose now that $\phi$ is an analytic linearization transformation for (3.1). As the reader may have already noticed, the functional equation (3.32) is a special case of (3.26). The Hilbert space equation (3.29) corresponding to (3.32) takes the form

$$M^\dagger|\phi\rangle = \lambda|\phi\rangle, \tag{3.35}$$

where $|\phi\rangle = \phi(a^\dagger)|0\rangle$.

It is clear that the linearization transformation can be recovered from the solution to (3.35) by

$$\phi(x) = \langle x|\phi\rangle \exp(\tfrac{1}{2}x^2),$$

where $|x\rangle$ is a normalized coherent state.

We have thus shown that the problem of finding analytic linearization transformations for nonlinear difference equations (3.1) can be brought down to the solution of the abstract Hilbert space eigenvalue equation (3.35). We note in addition that equations (3.30) corresponding to the occupation number representation for the abstract equation (3.35) can be written as

$$\phi(f_0) = \lambda\phi_0, \tag{3.36a}$$

$$\sum_{i=1}^{k} P_{ik}[f'_0, \ldots, f_0^{(k-i+1)}]\phi^{(i)}(f_0) = \lambda\phi_0^{(k)}, \qquad k = 1, 2, \ldots, \infty, \tag{3.36b}$$

where $P_{ik}$ are given by (3.31).

EXAMPLE. Consider the logistic equation (3.9). Assuming $\phi_n = \langle n|\phi\rangle$, where $|\phi\rangle$ fulfils (3.35), to be the solution of the linear first-order difference equation we make the following ansatz:

$$[g(N)a + h(N)]|\phi\rangle = |0\rangle, \tag{3.37}$$

where $N = a^\dagger a$ is the number operator; $g$ and $h$ are clearly analytic in $N$. Now equations (3.9) and (3.32) taken together yield $\langle 0|\phi\rangle = 0$. Hence, putting $\langle 1|\phi\rangle = 1$ we get $g(0) = 1$. Applying $a^\dagger$ to both sides of (3.37) we obtain the following equivalent form of the ansatz (3.37):

$$[\tilde{g}(N)a^\dagger + \tilde{h}(N)]|\phi\rangle = |1\rangle, \tag{3.38}$$

where $\tilde{h}(1) = 1$. On setting $g(N) \equiv 1$ and $h(N) = \nu N$ and taking into account (C.2a), (C.32) and (3.32) we arrive at the following linearization transformation:

$$\phi(x) = -\frac{1}{2}\ln(1-2x)$$

reducing the solution of (3.9), where $\mu = 2$, to the solution of its linear part. Hence, using (3.33) and (3.34) we obtain the solution of (3.9), where $\mu = 2$, such that

$$x_n(x_0) = \frac{1}{2}\left[1-(1-2x_0)^{2^n}\right].$$

An alternative method for the solution of (3.9) is to apply the linear transformation

$$\psi(x) = 1 - 2x,$$

converting (3.9), where $\mu = 2$, into the equation $x_{n+1} = x_n^2$. The celebrated solution to (3.9) for $\mu = 4$, can be recovered with the help of the actual treatment by involving in (3.37) the terms quadratic in $N$. More precisely, one should set $g(N) = \alpha N + 1$, and $h(N) = \beta N^2$. □

REMARK 1. The scheme of linearization in Hilbert space based on (3.35) can be treated as a generalization of the classical method of variation of constants to the case of difference equations. In fact, eq. (3.35) has the structure analogous to the equations arising in the linearization of nonlinear ordinary and partial differential equations (see section 4.2) which are shown to generalize the method of variation of constants. ■

REMARK 2. It is a natural question to ask whether one could find the linearization transformations for eq. (3.1) via (3.36). An experience with the logistic equation (3.9) when (3.36) imply the relations of the form $\phi_0^{(n+1)} = c_n(\mu)\phi_0^{(n)}$, with unknown general form of $c_n(\mu)$ for arbitrary $n$, indicates the negative answer. ■

The results of this section are applied in section 4.4 to study the Feigenbaum-Cvitanovic renormalization equations.

CHAPTER 4

# APPLICATIONS

## 4.1 First integrals

### *4.1.1 Ordinary differential equations*

Our aim in this section is to introduce the method for finding first integrals for nonlinear systems of ordinary differential equations [12]. Consider the real, analytic, autonomous system

$$\frac{d\mathbf{x}}{dt} = \mathbf{F}(\mathbf{x}). \tag{4.1}$$

As we have already seen (see section 1.3), the problem of determining the time-dependent first integrals for (4.1) that are analytic in spatial variables can be reduced to the solution of the following abstract evolution equation in Hilbert space:

$$\frac{d}{dt}|\phi, t\rangle = -M^{\dagger}|\phi, t\rangle, \tag{4.2}$$

where $M^{\dagger} = \mathbf{F}(\mathbf{a}^{\dagger})\cdot\mathbf{a}$ is the Hermitian conjugate of the "Hamiltonian" corresponding to (4.1).

Recall that the first integrals are linked to the solutions of (4.2) by

$$I(\mathbf{x}, t) = \langle\mathbf{x}|\phi, t\rangle e^{\frac{1}{2}\mathbf{x}^2}, \tag{4.3}$$

where $|\mathbf{x}\rangle$ is a normalized coherent state.

Consider equation (4.2). We seek the solutions to (4.2) of the form

$$|\phi, t\rangle = \alpha(t)|\phi\rangle, \tag{4.4}$$

that is, we apply the method of separation of variables. Substituting (4.4) into (4.2) we find

$$\alpha(t) = e^{-\lambda t}, \tag{4.5}$$
$$M^{\dagger}|\phi\rangle = \lambda|\phi\rangle, \tag{4.6}$$

where $\lambda$ is a separation constant.

On taking into account (4.3), (4.4) and (4.5) we arrive at the first integrals such that [33]:

$$I(\mathbf{x},t) = \tilde{\phi}(\mathbf{x})e^{-\lambda t}, \tag{4.7}$$

where $\tilde{\phi}(\mathbf{x}) = \langle \mathbf{x}|\phi\rangle e^{\frac{1}{2}\mathbf{x}^2}$ is analytic in $\mathbf{x}$ (see (C.31)).

It thus appears that the problem of finding the first integrals of the form (4.7), with analytic $\mathbf{x}$-dependent part, is equivalent to solving an abstract Hilbert space eigenvalue equation (4.6).

REMARK. We note that the operator $M^\dagger$ is Hermitian only in the case of the linear system (4.1) with Hermitian matrix. Therefore, the eigenvalue $\lambda$ can be in general complex. ■

We now specialize to the case of $\tilde{\phi}$ polynomial in $\mathbf{x}$. Taking into account (C.25) we find that the vector $|\phi\rangle$ can be then written as

$$|\phi\rangle = \sum_{\mathbf{n}\in\mathbf{S}} \alpha_{\mathbf{n}}|\mathbf{n}\rangle, \tag{4.8}$$

where the vectors $|\mathbf{n}\rangle$, $\mathbf{n} \in \mathbf{Z}_+^k$, span the occupation number representation and $\mathbf{S}$ is a bounded subset of $\mathbf{Z}_+^k$.

Suppose that an analytic function $f$ satisfies the system of equations $f(\mathbf{n}) = \lambda$, where $\mathbf{n} \in \mathbf{S}$. As is readily verified, the vector $|\phi\rangle$ given by (4.8) is then an eigenvector of the operator $f(\mathbf{N})$, where $N_i$, $i = 1, \ldots, k$, are the number operators, corresponding to the eigenvalue $\lambda$. Hence, using (4.6) we obtain

$$[M^\dagger, f(\mathbf{N})]|\phi\rangle = 0. \tag{4.9}$$

The general relation (4.9) is still too complicated to simplify the solution of (4.6). An even further simplification can be made by restricting to the operators $f$ linear in $\mathbf{N}$, that is we put in (4.9) $f(\mathbf{N}) = \mathbf{c}\cdot\mathbf{N}$. In other words, we set $|\phi\rangle$ to be the common eigenvector of the operator $M^\dagger$ and the linear combination $\mathbf{c}\cdot\mathbf{N}$ of the number operators $N_i$, $i = 1, \ldots, k$, i.e.,

$$[M^\dagger, \mathbf{c}\cdot\mathbf{N}]|\phi\rangle = 0. \tag{4.10}$$

It appears that there exist nontrivial cases when the relations (4.6) and (4.10) allow to determine in a simple way the first integrals. We now illustrate this observation by the following examples.

EXAMPLE 1. Consider the Lorenz system

$$\begin{aligned}
\frac{dx_1}{dt} &= \sigma(x_2 - x_1), \\
\frac{dx_2}{dt} &= -x_2 - x_1x_3 + rx_1, \\
\frac{dx_3}{dt} &= x_1x_2 - bx_3.
\end{aligned} \tag{4.11}$$

The conjugate $M^\dagger$ of the "Hamiltonian" corresponding to (4.11) can be written as

$$M^\dagger = -\sigma N_1 - N_2 - bN_3 + \sigma a_2^\dagger a_1 + r a_1^\dagger a_2 - a_1^\dagger a_3^\dagger a_2 + a_1^\dagger a_2^\dagger a_3.$$

Hence, taking into account (C.4) we obtain the following commutation relation:

$$[M^\dagger, \mathbf{c}\cdot\mathbf{N}] = \sigma(c_1 - c_2)a_2^\dagger a_1 + r(c_2 - c_1)a_1^\dagger a_2 + (c_1 + c_3 - c_2)a_1^\dagger a_3^\dagger a_2 + (c_3 - c_1 - c_2)a_1^\dagger a_2^\dagger a_3. \tag{4.12}$$

On putting $c_1 = c_2$, $c_3 = 2c_1$ we reduce (4.12) to

$$[M^\dagger, \mathbf{c}\cdot\mathbf{N}] = 2c_1 a_1^\dagger a_3^\dagger a_2. \tag{4.13}$$

The relations (4.13) and (4.10) taken together suggest the following ansatz:

$$|\phi\rangle = \sum_{n_1 n_3} \alpha_{n_1 n_3} |n_1 0 n_3\rangle. \tag{4.14}$$

In fact, inserting (4.14) into (4.6) we find easily

$$\lambda = -2\sigma, \qquad b = 2\sigma,$$
$$|\phi\rangle = \alpha_{20}|200\rangle + \alpha_{01}|001\rangle,$$

where $\sqrt{2}\,\sigma\alpha_{20} = -\alpha_{01}$. Hence setting $\alpha_{20} = \sqrt{2}$ and using (4.7) we finally obtain the first integral of the form

$$I_1(\mathbf{x}, t) = (x_1^2 - 2\sigma x_3)e^{2\sigma t}.$$

Analogously, for $c_3 = c_2$, $c_1 = 0$, $r = 0$ the relation (4.12) becomes

$$[M^\dagger, \mathbf{c}\cdot\mathbf{N}] = -\sigma c_2 a_2^\dagger a_1.$$

From (4.10) we deduce the ansatz

$$|\phi\rangle = \sum_{n_2 n_3} \alpha_{n_2 n_3} |0 n_2 n_3\rangle. \tag{4.15}$$

Substituting (4.15) into (4.6) yields

$$\lambda = -2, \qquad b = 1,$$
$$|\phi\rangle = \alpha_{20}|020\rangle + \alpha_{02}|002\rangle,$$

where $\alpha_{20} = \alpha_{02}$. Hence, taking into account (4.7) we get the following first integral:

$$I_2(\mathbf{x}, t) = (x_2^2 + x_3^2)e^{2t}.$$

Finally, putting $c_1 = c_2 = c_3$ we obtain from (4.12) the following commutation relation:

$$[M^\dagger, \mathbf{c}\cdot\mathbf{N}] = c_1 a_1^\dagger (a_3^\dagger a_2 - a_2^\dagger a_3).$$

It is easy to verify that (4.10) implies the ansatz

$$|\phi\rangle = \sum_n \alpha_n(|n20\rangle + |n02\rangle) + \sum_m \beta_m|m00\rangle. \tag{4.16}$$

Inserting (4.16) into (4.6) we find

$$\lambda = -2, \qquad \sigma = 1, \qquad b = 1,$$
$$|\phi\rangle = \alpha_0(|020\rangle + |002\rangle) + \beta_2|200\rangle,$$

where $\alpha_0 r = -\beta_2$. Hence, with the use of (4.7) the following first integral is obtained:

$$I_3(\mathbf{x}, t) = (-rx_1^2 + x_2^2 + x_3^2)e^{2t}.$$

The first integrals for the Lorenz system of fourth order in $\mathbf{x}$ [34] can be recovered with the help of the introduced method by involving in (4.9) the operators quadratic in $\mathbf{N}$. □

EXAMPLE 2. Consider the Rikitake two-disc dynamo system [35]:

$$\begin{aligned}\frac{dx_1}{dt} &= -\mu x_1 + x_2 x_3,\\ \frac{dx_2}{dt} &= -\mu x_2 - \alpha x_1 + x_1 x_3,\\ \frac{dx_3}{dt} &= 1 - x_1 x_2.\end{aligned} \tag{4.17}$$

The system (4.17) is known to show chaotic behaviour for a wide range of parameters $\mu$ and $\alpha$. The operator $M^\dagger$ corresponding to (4.17) is

$$M^\dagger = -\mu N_1 - \mu N_2 - \alpha a_1^\dagger a_2 + a_3 + a_2^\dagger a_3^\dagger a_1 + a_1^\dagger a_3^\dagger a_2 - a_1^\dagger a_2^\dagger a_3.$$

Therefore,

$$\begin{aligned}[M^\dagger, \mathbf{c}\cdot\mathbf{N}] &= \alpha(c_1 - c_2)a_1^\dagger a_2 + (c_1 - c_2 - c_3)a_2^\dagger a_3^\dagger a_1 + (c_2 - c_1 - c_3)a_1^\dagger a_3^\dagger a_2\\ &\quad + (c_1 + c_2 - c_3)a_1^\dagger a_2^\dagger a_3 + c_3 a_3.\end{aligned}$$

On setting $c_1 = c_2$, $c_3 = 0$ we arrive at the following commutation relation:

$$[M^\dagger, \mathbf{c}\cdot\mathbf{N}] = 2c_1 a_1^\dagger a_2^\dagger a_3.$$

Hence, substituting the ansatz

$$|\phi\rangle = \sum_{n_1 n_2} \alpha_{n_1 n_2}|n_1 n_2 0\rangle$$

into (4.6) we find easily

$$\lambda = -2\mu, \qquad \alpha = 0, \tag{4.18a}$$
$$|\phi\rangle = \alpha_{20}|200\rangle + \alpha_{02}|020\rangle, \tag{4.18b}$$

where $\alpha_{20} = -\alpha_{02}$. Taking into account (4.18) and (4.7) we get the first integral for (4.17) such that

$$I(\mathbf{x}, t) = (x_1^2 - x_2^2)e^{2\mu t}. \qquad \square$$

*4.1.2 Partial differential equations*

We now introduce a method for finding first integrals for nonlinear partial differential equations of the evolution type [13]. Consider, for simplicity, the one-dimensional real analytic system (2.11) of the form

$$\partial_t u(x,t) = F(u, \partial_x u, \ldots, \partial_x^r u), \tag{4.19}$$

where $u$: $\mathbf{R} \times \mathbf{R} \to \mathbf{R}$, $F$ is analytic in $u$, $\partial_x u$, $\ldots$, $\partial_x^r u$, and $u \in L^2_{\mathbf{R}}(\mathbf{R}, dx)$.

We recall (see section 2.3) that the problem of determining analytic first integrals for (4.19) can be brought down to the solution of an abstract Hilbert space eigenvalue equation

$$M^\dagger|\phi\rangle = 0, \tag{4.20}$$

where $M^\dagger = \int dx\, F(a^\dagger(x), a^{\dagger\prime}(x), \ldots, a^{\dagger(r)}(x))a(x)$ is the Hermitian conjugate of the "Hamiltonian" corresponding to (4.19). Here $a^{\dagger\prime}(x) \equiv da^\dagger(x)/dx$.

The first integrals $I[u]$ for (4.19) can be recovered from solutions to (4.20) by

$$I[u] = \langle u|\phi\rangle \exp\left(\frac{1}{2}\int dx\, u^2\right), \tag{4.21}$$

where $|u\rangle$ is a normalized functional coherent state.

Writing (4.20) in the coordinate representation (see appendix C) and performing a Fourier transformation we obtain a linear system of algebraic equations. It turns out that there exist nontrivial cases when the first integrals can be found easily by solving such linear equations. For an easy illustration of this observation we now rederive the first integrals for the Korteweg-de Vries equation.

EXAMPLE. Consider the Korteweg-de Vries equation

$$\partial_t u = -\partial_x^3 u + 6u\partial_x u. \tag{4.22}$$

The conjugate $M^\dagger$ of the "Hamiltonian" corresponding to (4.22) is given by

$$M^\dagger = \int dx\, (-a^{\dagger\prime\prime\prime}(x) + 6a^{\dagger\prime}(x)a^\dagger(x))a(x).$$

On writing (4.20) in the coordinate representation (see formulae (C.57)) we arrive at the linear system of hierarchy equations such that

$$\partial_{x_1}^3 \phi_1(x_1) = 0, \tag{4.23a}$$

$$6\sum_{i=1}^{n+1}\sum_{j\neq i} \partial_{x_j}\delta(x_i - x_j)\phi_n(x_1, \ldots, \check{x}_i, \ldots, x_{n+1}) + \sum_{i=1}^{n+1} \partial_{x_i}^3 \phi_{n+1}(x_1, \ldots, x_{n+1}) = 0, \qquad n = 1, 2, \ldots, \infty, \tag{4.23b}$$

where $\phi_n(x_1,\ldots,x_n) = \langle x_1,\ldots,x_n|\phi\rangle$ and the reversed hat over $x_i$ designates that this variable should be omitted from the set $\{x_1,\ldots,x_{n+1}\}$. We note that the solution of (4.20) and the solution of (4.23) are related by

$$|\phi\rangle = \sum_{n=1}^{\infty} \frac{1}{n!} \int dx_1 \cdots dx_n \, \phi_n(x_1,\ldots,x_n)|x_1,\ldots,x_n\rangle. \tag{4.24}$$

It should also be noted that the functions $\phi_n(x_1,\ldots,x_n)$ are symmetric in $x_1, \ldots, x_n$ (see (C.54)). On performing the Fourier transformation we obtain from (4.23) the system of linear algebraic equations of the form

$$k_1^3\tilde{\phi}_1(k_1) = 0, \tag{4.25a}$$

$$6\sum_{i=1}^{n+1}\sum_{j\neq i} k_i\tilde{\phi}_n(k_1,\ldots,k_j,\ldots,k_{n+1},\check{k}_i)|_{k_j\to k_i+k_j} + \sum_{i=1}^{n+1} k_i^3\tilde{\phi}_{n+1}(k_1,\ldots,k_{n+1}) = 0, \qquad n = 1,\, 2,\, \ldots,\, \infty. \tag{4.25b}$$

Using the identity

$$\sum_{i=1}^{n+1}\sum_{j\neq i} k_i\psi_n(k_1,\ldots,k_j,\ldots,k_{n+1},\check{k}_i)|_{k_j\to k_i+k_j} = -2\sum_{i=1}^{n+1} k_i^{2s+1}\delta\left(\sum_{j=1}^{n+1} k_j\right),$$

where $\psi_n = \sum_{i=1}^{n} k_i^{2s}\delta\left(\sum_{j=1}^{n} k_j\right)$, $s = 0,\, 1,\, 2,\, \ldots$, the following solutions to (4.25) can be found easily:

$$\tilde{\phi}_1 = \delta(k_1), \qquad \tilde{\phi}_n = 0, \quad n \geq 2, \tag{4.26a}$$

$$\tilde{\phi}_1 = 0, \qquad \tilde{\phi}_2 = \delta(k_1+k_2), \qquad \tilde{\phi}_n = 0, \quad n \geq 3, \tag{4.26b}$$

$$\tilde{\phi}_1 = 0, \qquad \tilde{\phi}_2 = (k_1^2+k_2^2)\delta(k_1+k_2), \qquad \tilde{\phi}_3 = 12\delta(k_1+k_2+k_3), \qquad \tilde{\phi}_n = 0, \quad n \geq 4, \tag{4.26c}$$

$$\tilde{\phi}_1 = 0, \qquad \tilde{\phi}_2 = (k_1^4+k_2^4)\delta(k_1+k_2), \qquad \tilde{\phi}_3 = 10(k_1^2+k_2^2+k_3^2)\delta(k_1+k_2+k_3), \qquad \tilde{\phi}_4 = 120\delta(k_1+k_2+k_3+k_4), \qquad \tilde{\phi}_n = 0, \quad n \geq 5, \tag{4.26d}$$

$$\tilde{\phi}_1 = 0, \qquad \tilde{\phi}_2 = (k_1^6+k_2^6)\delta(k_1+k_2), \qquad \tilde{\phi}_3 = 14(k_1^4+k_2^4+k_3^4)\delta(k_1+k_2+k_3), \qquad \tilde{\phi}_4 = 140(k_1^2+k_2^2+k_3^2+k_4^2)\delta(k_1+k_2+k_3+k_4), \qquad \tilde{\phi}_5 = 1680\delta(k_1+k_2+k_3+k_4+k_5), \qquad \tilde{\phi}_n = 0, \quad n \geq 6. \tag{4.26e}$$

We observe that the solutions (4.26b)–(4.26e) are generated by the function $\tilde{\phi}_2 = (k_1^{2m}+k_2^{2m})\delta(k_1+k_2)$, $m = 0,\, 1,\, 2,\, 3$. It is suggested that every remaining element of the infinite set of solutions to (4.25), corresponding to the infinite number of polynomial integrals to the Korteweg-de Vries equation is generated in the same manner. On taking the Fourier's inverse transformations to (4.26a)–(4.26e) and using (4.24), (4.21) and (C.63) we obtain the following first integrals for the Korteweg-de Vries equation:

$$
\begin{aligned}
I_1[u] &= \int dx\, u, \\
I_2[u] &= \int dx\, u^2, \\
I_3[u] &= \int dx\, (-u\partial_x^2 u + 2u^3), \\
I_4[u] &= \int dx\, (u\partial_x^4 u - 5u^2\partial_x^2 u + 5u^4), \\
I_5[u] &= \int dx\, (-u\partial_x^6 u + 7u^2\partial_x^4 u - (70/3)u^3\partial_x^2 u + 14u^5). \qquad \Box
\end{aligned}
$$

What about the generation of infinite sequence of first integrals? It seems that the simplest possibility is to seek the operator $\mathcal{R}$ such that the linear recurrence in Hilbert space

$$|i+1\rangle = \mathcal{R}|i\rangle, \qquad i = 1, 2, \ldots, \tag{4.27}$$

where

$$M^\dagger|i\rangle = 0, \qquad i = 1, 2, \ldots, \tag{4.28}$$

generates the infinite sequence of first integrals for (4.19) (whenever they exist) via (4.21), i.e. the following relations hold:

$$I_i[u] = \langle u|i\rangle \exp\left(\frac{1}{2}\int dx\, u^2\right), \qquad i = 1, 2, \ldots, . \tag{4.29}$$

With this remark serving as our motivation we now proceed to study the master symmetries for partial differential equations within the Hilbert space formalism introduced in section 2.3 [31]. We begin by recalling that if (4.19) possesses an infinite number of symmetries $\sigma_i$, $i = 1, 2, \ldots$, then a function $\tau$ satisfying

$$\sigma_{i+1} = [\sigma_i, \tau], \qquad i = 1, 2, \ldots, \tag{4.30}$$

where the bracket is given by (2.39),

is called the *master symmetry* of (4.19) [36]. The recursive scheme (4.30) for obtaining infinitely many symmetries is called the $\tau$-scheme. On using the isomorphism (2.41) we find that the $\tau$-scheme (4.30) is represented within the Hilbert space approach by the following operator recurrence:

$$\Sigma_{i+1} = [\Sigma_i, R], \qquad i = 1, 2, \ldots, \tag{4.31}$$

where $\Sigma_i = \int dx\, a^\dagger\sigma_i(a)$ and $R = \int dx\, a^\dagger\tau(a)$.

It thus appears that the boson operator $R$ provides a Hilbert space recursive procedure for generation of symmetries. We now demonstrate by the example of the Korteweg-de Vries equation that the Hermitian conjugate of the operator $R$ such that

$$R^\dagger = \int dx\, \tau(a^\dagger)a,$$

is nothing but the operator $\mathcal{R}$ defined by (4.27), that is the operator $R^\dagger$ generates the first integrals via (4.29) and

$$|i+1\rangle = R^\dagger |i\rangle.$$

EXAMPLE. Consider the Korteweg-de Vries equation (4.22). This equation is well-known to possess infinitely many polynomial symmetries. The corresponding $\tau$-scheme can be written as

$$\sigma_{i+1} = [\sigma_i, \tau], \qquad \sigma_1 = \partial_x u, \qquad i = 1, 2, \ldots, \infty,$$

where the master symmetry $\tau$ is [36]:

$$\tau = x(-\partial_x^3 u + 6u\partial_x u) - 4\partial_x^2 u + 2\partial_x u \partial_x^{-1} u + 8u^2. \tag{4.32}$$

Hence the operator recurrence (4.31) becomes

$$\Sigma_{i+1} = [\Sigma_i, R], \qquad \Sigma_1 = \int dx\, a^\dagger a', \qquad i = 1, 2, \ldots, \infty, \tag{4.33}$$

where $\Sigma_i = \int dx\, a^\dagger \sigma_i(a)$ and $R = \int dx\, a^\dagger \tau(a)$. By virtue of (4.32) the Hermitian conjugate of the operator $R$ is given by

$$R^\dagger = \int dx\, [x(-a^{\dagger\prime\prime\prime} + 6a^\dagger a^{\dagger\prime}) - 4a^{\dagger\prime\prime} + 2a^{\dagger\prime} a^{\dagger -\prime} + 8a^{\dagger 2}]a,$$

where $a^{\dagger-\prime}(x) \equiv (d/dx)^{-1} a^\dagger(x)$.

We now introduce a vector of the form

$$|\psi\rangle = -\int x|x\rangle\, dx,$$

where the vectors $|x\rangle$ span the one-particle subspace of the coordinate representation (see appendix C).

The reader can easily check that this vector satisfies

$$R^\dagger |\psi\rangle = 0.$$

Hence, defining the vectors such that

$$|i\rangle = \Sigma_i^\dagger |\psi\rangle, \qquad i = 1, 2, \ldots, \infty, \tag{4.34}$$

and taking the Hermitian conjugate of (4.33) we obtain the following recurrence in Hilbert space:

$$|i+1\rangle = R^\dagger|i\rangle, \qquad |1\rangle = \int dx\,|x\rangle, \qquad i = 1, 2, \dots, \infty. \tag{4.35}$$

We now prove that the infinite sequence of first integrals for the Korteweg-de Vries equation is generated by (4.35) and (4.29). We first show that (4.28) holds, that is the functionals given by (4.29) are the first integrals. An easy inspection using (4.34), (2.50) and the Hermitian conjugate of (4.33) shows that (4.28) is equivalent to

$$\Sigma_i^\dagger|1\rangle = 0, \qquad i = 1, 2, \dots, \infty.$$

Furthermore, the symmetries for the Korteweg-de Vries equation are well-known to be the spatial derivatives of functional derivatives (gradients) of first integrals. Therefore,

$$\langle u|\Sigma_i^\dagger|1\rangle = e^{-\frac{1}{2}\int dx\, u^2}\int dx\,\sigma_i(u) = 0, \qquad i = 1, 2, \dots, \infty,$$

where $|u\rangle$ is a normalized functional coherent state and $u$ fulfils (4.22). We have thus proved that (4.28) is valid.

We now demonstrate that we have infinitely many first integrals generated by (4.35) and (4.29). We first prove that $I_i \neq I_j$ for $i \neq j$. This means that there are no vectors $|i_*\rangle$, where $i_* \in \mathbf{N}$, which obey

$$R^\dagger|i_*\rangle = |i_*\rangle. \tag{4.36}$$

Consider the following boson operator corresponding via the isomorphism (2.41) to the scaling symmetry [36] of (4.22):

$$Y = \int dx\, a^\dagger(\tfrac{1}{2}xa' + a).$$

We have,

$$[R, Y] = R, \tag{4.37}$$
$$[\Sigma_i, Y] = (i - \tfrac{1}{2})\Sigma_i, \qquad i = 1, 2, \dots, \infty, \tag{4.38}$$

the equation (4.38) following from (4.37), (4.33) and the Jacobi identity. Further, it is easy to verify that the Hermitian conjugate $Y^\dagger$ of the operator $Y$ satisfies

$$Y^\dagger|\psi\rangle = 0.$$

Hence, taking the conjugate of (4.38) and using (4.34) we get

$$Y^\dagger|i\rangle = (i - \tfrac{1}{2})|i\rangle, \qquad i = 1, 2, \dots, \infty. \tag{4.39}$$

Let us assume now that (4.36) is valid. Applying $R^\dagger$ to both sides of (4.39) and taking the conjugate of (4.37) we conclude that $i_* = i_* - 1$. This contradiction completes the proof.

It remains to show that none of first integrals $I_i$, $i = 1, 2, \ldots$, vanishes. Clearly, the vanishing of a first integral implies that

$$R^\dagger|j\rangle = 0, \tag{4.40}$$

with fixed $j \in \mathbf{N}$.

Suppose that we are given a boson operator $Z$ such that

$$Z = \int dx\, a^\dagger.$$

This operator satisfies

$$\begin{aligned} [R, Z] &= 16Y, \\ [\Sigma_i, Z] &= c_i \Sigma_{i-1}, \qquad i = 2, 3, \ldots, \infty, \end{aligned} \tag{4.41}$$

where $c_i = 6 + 8i(i-2)$.

It follows that

$$Z^\dagger|i\rangle = c_i|i-1\rangle, \qquad i = 2, 3, \ldots, \infty. \tag{4.42}$$

Suppose now that (4.40) holds true. A straightforward calculation shows that (4.40) is not valid for $j = 1$. Thus, we can put $j \geq 2$. The conjugate of (4.41) and eqs. (4.42) and (4.39) taken together yield $j = \pm\frac{1}{2}$, a contradiction.

We have thus proved that the operator $R^\dagger$ generates via (4.35) and (4.29) the infinite hierarchy of first integrals for the Korteweg-de Vries equation.

REMARK. It should be noted that the scheme of generating first integrals provided by the operator $R^\dagger$ is not restricted to the case of completely integrable equations. Indeed, the operator $R^\dagger$ works also in the case with equations possessing infinitely many symmetries and finite number of first integrals. As an illustration of this observation, we consider the Burgers equation

$$\partial_t u = \partial_x^2 u - u\partial_x u. \tag{4.43}$$

The operator recurrence in Hilbert space (4.31) corresponding to the $\tau$-scheme for generating the infinite sequence of symmetries for the Burgers equation [36] is of the form

$$\Sigma_{i+1} = [\Sigma_i, R], \qquad \Sigma_1 = \int dx\, a^\dagger a', \qquad i = 1, 2, \ldots, \infty,$$

where the Hilbert space counterpart $R$ of the master symmetry $\tau$ is

$$R = \int dx\, a^\dagger[x(a'' - aa') - \tfrac{1}{2}a^2].$$

As with the Korteweg-de Vries equation, we arrive at the following recurrence:

$$|i+1\rangle = R^{\dagger}|i\rangle, \qquad |1\rangle = \int dx\,|x\rangle, \qquad i = 1, 2, \ldots, \infty. \tag{4.44}$$

The reader can easily check that the following relation holds:

$$R^{\dagger}|1\rangle = 0.$$

We have thus shown that the first integral for (4.43) such that

$$I_1[u] = \langle u|1\rangle \exp\left(\frac{1}{2}\int dx\, u^2\right) = \int dx\, u, \tag{4.45}$$

is the only one generated by (4.44) and (4.29). We note at the same time that (4.45) is well-known to be the only first integral for the Burgers equation. ∎

## 4.2 Linearization transformations

### *4.2.1 Ordinary differential equations*

This section discusses the linearization transformations for ordinary differential equations [37]. Consider the following real autonomous system:

$$\frac{d\mathbf{x}}{dt} = \mathbf{F}(\mathbf{x}), \qquad \mathbf{x}(0) = \mathbf{x}_0, \tag{4.46}$$

where $\mathbf{F}$: $\mathbf{R}^k \to \mathbf{R}^k$ is analytic in $\mathbf{x}$.

We now examine the transformation of variables within the Hilbert space formalism introduced in chapter 1. We begin by recalling that the "Hamiltonian" $M$ corresponding to (4.46) is

$$M = \mathbf{a}^{\dagger}\cdot\mathbf{F}(\mathbf{a}). \tag{4.47}$$

Consider the following mapping:

$$\mathbf{x}' = \boldsymbol{\phi}(\mathbf{x}), \tag{4.48}$$

where $\mathbf{x}$ fulfils (4.46) and $\boldsymbol{\phi}$ is analytic in $\mathbf{x}$.

Taking into account (1.2b) and (C.2a) we find that under (4.48) the "Hamiltonian" $M$ transforms as follows:

$$M' = \mathbf{a}^{\dagger}\cdot[\boldsymbol{\phi}(\mathbf{a}), M]. \tag{4.49}$$

Suppose now that (4.48) is a linearization mapping. This means that (4.48) converts (4.46) into the linear system such that

$$\frac{d\mathbf{x}'}{dt} = L\mathbf{x}', \tag{4.50}$$

where $L$: $\mathbf{R}^k \to \mathbf{R}^k$ is a linear operator.

It follows from (4.49) that the linearization mapping satisfies the following commutation relation:

$$[\phi(\mathbf{a}), M] = L\phi(\mathbf{a}). \tag{4.51}$$

On introducing the vectors $|\phi_i\rangle$, $i = 1, \ldots, k$, of the form

$$|\phi_i\rangle = \phi_i(\mathbf{a}^\dagger)|\mathbf{0}\rangle, \qquad i = 1, \ldots, k, \tag{4.52}$$

and taking the Hermitian conjugate of (4.51) we obtain the following system of linear equations in Hilbert space:

$$M^\dagger|\phi_i\rangle = \sum_{j=1}^{k} L_{ij}|\phi_j\rangle, \qquad i = 1, \ldots, k, \tag{4.53}$$

where $L_{ij}$ is the matrix corresponding to the operator $L$.

Using (4.52) and (C.31) we find that the linearization mapping can be recovered from the solution to (4.53) by

$$\phi_i(\mathbf{x}) = \langle\mathbf{x}|\phi_i\rangle e^{\frac{1}{2}\mathbf{x}^2}, \qquad i = 1, \ldots, k, \tag{4.54}$$

where $|\mathbf{x}\rangle$ is a normalized coherent state.

It thus appears that the problem of determining the linearization transformation for the nonlinear dynamical system (4.46) can be brought down to the solution of the system of linear equations in Hilbert space (4.53).

EXAMPLE. Consider the Riccati system

$$\frac{d\mathbf{x}}{dt} = L\mathbf{x} + (\mathbf{c}\cdot\mathbf{x})\mathbf{x}, \qquad \mathbf{x}(0) = \mathbf{x}_0, \tag{4.55}$$

where $L\colon \mathbf{R}^k \to \mathbf{R}^k$ is a linear operator. The "Hamiltonian" (4.47) corresponding to (4.55) takes the form

$$M = \mathbf{a}^\dagger\cdot L\mathbf{a} + N\mathbf{c}\cdot\mathbf{a}, \tag{4.56}$$

where $N = \mathbf{a}^\dagger\cdot\mathbf{a}$ is the total number operator. Making use of (4.56) and (C.5) we arrive at the following symmetry of the "Hamiltonian" $M$:

$$[M, N(1 + \tilde{L}^{-1}\mathbf{c}\cdot\mathbf{a})] = 0, \tag{4.57}$$

where $\tilde{L}$ designates the transpose of the operator $L$. Taking into account (4.51), (4.57) and the Jacobi identity we get

$$[\phi_i(\mathbf{a}), N(1 + \tilde{L}^{-1}\mathbf{c}\cdot\mathbf{a})] = \phi_i(\mathbf{a}), \qquad i = 1, \ldots, k.$$

Proceeding analogously as in the case of the commutation relation (4.51) we find

$$(1 + \tilde{L}^{-1}\mathbf{c}\cdot\mathbf{a}^\dagger)N|\phi_i\rangle = |\phi_i\rangle, \qquad i = 1, \ldots, k. \tag{4.58}$$

Equations (4.58) and (C.5) taken together yield

$$(N-1)(1+\tilde{L}^{-1}\mathbf{c}\cdot\mathbf{a}^\dagger)|\phi_i\rangle = 0, \qquad i=1,\ldots,k.$$

Recalling that there is no nontrivial closed subspace of the Hilbert space, where the Bose operators act, which is invariant under their action, we can write

$$(N-1)|\psi_i\rangle = 0, \qquad i=1,\ldots,k, \tag{4.59}$$

where

$$|\psi_i\rangle = (1+\tilde{L}^{-1}\mathbf{c}\cdot\mathbf{a}^\dagger)|\phi_i\rangle \tag{4.60}$$

and $|\psi_i\rangle = 0$ if and only if $|\phi_i\rangle = 0$. The following solution to (4.59) can be easily obtained:

$$|\psi_i\rangle = |\mathbf{e}_i\rangle = a_i^\dagger|\mathbf{0}\rangle, \qquad i=1,\ldots,k, \tag{4.61}$$

where $|\mathbf{e}_i\rangle$ is a basis vector of the occupation number representation and $\mathbf{e}_i$ is a unit vector of $\mathbf{R}^k$ On using (4.60) and (4.61) we arrive at the following solution of (4.53):

$$|\phi_i\rangle = \frac{a_i^\dagger}{1+\tilde{L}^{-1}\mathbf{c}\cdot\mathbf{a}^\dagger}|\mathbf{0}\rangle, \qquad i=1,\ldots,k.$$

Hence, with the use of (4.54) we finally get the linearization transformation for the Riccati system (4.55) such that

$$\phi(\mathbf{x}) = \frac{\mathbf{x}}{1+\tilde{L}^{-1}\mathbf{c}\cdot\mathbf{x}}. \tag{4.62}$$

On taking the inverse transformation of (4.62) and using (4.50) we find that the solution to (4.55) can be written as

$$\mathbf{x} = \frac{\mathbf{x}'}{1-\tilde{L}^{-1}\mathbf{c}\cdot\mathbf{x}'}, \tag{4.63}$$

where $\mathbf{x}' = \phi(\mathbf{x})$ is the solution of the linear part of the system (4.55) subject to the initial condition

$$\mathbf{x}_0' = \mathbf{x}_0/F,$$

where $F = 1+\tilde{L}^{-1}\mathbf{c}\cdot\mathbf{x}_0$.

As mentioned in ref. [38], the singularity of these initial data arising when $F=0$, can be easily removed by multiplying the numerator and denominator of (4.63) by the factor $F$. Consequently, the solution $\mathbf{x}(\mathbf{x}_0, t)$ to (4.55) is of the form

$$\mathbf{x}(\mathbf{x}_0,t) = \frac{\mathbf{v}(\mathbf{x}_0,t)}{1-\tilde{L}^{-1}\mathbf{c}\cdot(\mathbf{v}(\mathbf{x}_0,t)-\mathbf{x}_0)},$$

where $\mathbf{v}(\mathbf{x}_0,t)$ is the solution to the linear part of the system (4.55) subject to the initial condition $\mathbf{x}(0)=\mathbf{x}_0$. □

*4.2.2 Partial differential equations*

In this section we introduce a method for the linearization of nonlinear partial differential equations [37]. Consider the analytic equation

$$\partial_t u(x,t) = F(u, D^\alpha u), \qquad u(x,0) = u_0(x), \tag{4.64}$$

where $u$: $\mathbf{R}^s \times \mathbf{R} \to \mathbf{R}$, $F$ is analytic in $u$, $D^\alpha u$ and $u_0 \in L^2_{\mathbf{R}}(\mathbf{R}^s, d^s x)$.

We now use the Hilbert space formulation for partial differential equations developed in chapter 2 to study the transformations of variables. We first recall that the "Hamiltonian" corresponding to (4.64) is of the form

$$M = \int d^s x\, a^\dagger(x) F(a(x), D^\alpha a(x)).$$

Consider the analytic mapping

$$u' = \phi[u|x],$$

where $\phi$ is analytic in $u$.

Proceeding analogously as in the case with ordinary differential equations discussed in previous section we find that the "Hamiltonian" $M$ obeys the transformation law

$$M' = \int d^s x\, a^\dagger(x)[\phi[a|x], M].$$

We now specialize to the case of the linearization transformations such that

$$\partial_t u' = Lu', \tag{4.65}$$

where $L$ is a linear differential operator.

As with ordinary differential equations, we arrive at the following relation satisfied by the linearization mapping:

$$[\phi[a|x], M] = L\phi[a|x],$$

which leads to

$$M^\dagger|\phi(x)\rangle = L|\phi(x)\rangle, \tag{4.66}$$

where $|\phi(x)\rangle = \phi[a^\dagger|x]|0\rangle$.

It is clear that the linearization transformation and the solution to (4.66) are related by

$$\phi[u|x] = \langle u|\phi(x)\rangle \exp\left(\frac{1}{2}\int d^s x\, u^2\right), \tag{4.67}$$

where $|u\rangle$ is a normalized functional coherent state.

We have thus shown that the problem of finding the linearization transformations for nonlinear partial equations of the evolution type (4.64) can be brought down to the solution of the abstract, linear equation in Hilbert space (4.66).

EXAMPLE. Consider the Burgers equation

$$\partial_t u = \nu \partial_x^2 u - u\partial_x u, \qquad u(x,0) = u_0(x) \in L^2_{\mathbf{R}}(\mathbf{R}, dx). \tag{4.68}$$

The Hermitian conjugate of the "Hamiltonian" corresponding to (4.68) is given by

$$M^\dagger = \nu \int dx\, a^{\dagger\prime\prime}(x)a(x) - \int dx\, a^\dagger(x)a^{\dagger\prime}(x)a(x).$$

Let $L = \nu\partial_x^2$. The equation (4.66) becomes

$$M^\dagger|\phi(x)\rangle = \nu\partial_x^2|\phi(x)\rangle. \tag{4.69}$$

The reader may notice that the solution of (4.69) is determined up to the additive factor $(ax+b)|0\rangle$, where $a$, $b$ are arbitrary constants. Writing (4.69) in the coordinate representation we obtain the following linear system of hierarchy equations:

$$\begin{aligned}
&\partial_{x_1}^2 \phi_1(x;x_1) = \partial_x^2\phi_1(x;x_1),\\
&\sum_{r=1}^{n+1}\sum_{s\neq r} \partial_{x_s}\delta(x_r - x_s)\phi_n(x;x_1,\ldots,\check{x}_r,\ldots,x_{n+1})\\
&\qquad = \nu\left(\partial_x^2 - \sum_{r=1}^{n+1}\partial_{x_r}^2\right)\phi_{n+1}(x;x_1,\ldots,x_{n+1}), \qquad n = 1, 2, \ldots, \infty,
\end{aligned}$$

where $\phi_n(x;,x_1,\ldots,x_n) = \langle x_1,\ldots,x_n|\phi(x)\rangle$ and the reversed hat over $x_r$ denotes that this variable should be omitted from the set $\{x_1,\ldots,x_{n+1}\}$. Hence, passing to the Fourier transformation we arrive at the system of linear algebraic equations such that

$$(k^2 - k_1^2)\tilde{\phi}_1(k;k_1) = 0, \tag{4.70a}$$

$$\begin{aligned}
&i\sum_{r=1}^{n+1}\sum_{s\neq r} k_r\tilde{\phi}_n(k;k_1,\ldots,k_s,\ldots,k_{n+1},\check{k}_r)|_{k_s\to k_r+k_s}\\
&\qquad = \nu\left(k^2 - \sum_{r=1}^{n+1}k_r^2\right)\tilde{\phi}_{n+1}(k;k_1,\ldots,k_{n+1}), \qquad n = 1, 2, \ldots, \infty.
\end{aligned} \tag{4.70b}$$

Making use of the identities

$$\begin{aligned}
&(k^2 - k_1^2 - k_2^2)^{-1}\delta(k+k_1+k_2) = \frac{1}{2k_1k_2}\delta(k+k_1+k_2),\\
&\left[\left(k^2 - \sum_{r=1}^{n+1}k_r^2\right)^{-1}\sum_{i_1>\ldots>i_{n-1}}\frac{1}{k_{i_1}\cdots k_{i_{n-1}}}\right]\delta\left(k+\sum_{r=1}^{n+1}k_r\right)\\
&\qquad = \frac{1}{2}\left(\prod_{r=1}^{n+1}\frac{1}{k_r}\right)\delta\left(k+\sum_{r=1}^{n+1}k_r\right), \qquad i_1,\ldots,i_{n-1}\in\{1,\ldots,n+1\}, \qquad n\geq 2,
\end{aligned}$$

we find easily the following solution of (4.70):

$$\tilde{\phi}_n(k; k_1, \ldots, k_n) = \frac{1}{(-2\nu i)^n} \left( \prod_{j=1}^{n} \frac{1}{k_j} \right) \delta \left( k + \sum_{j=1}^{n} k_j \right) \tag{4.71}$$

On performing the Fourier's inverse transformations to (4.71) and using

$$|\phi(x)\rangle = |0\rangle + \sum_{n=1}^{\infty} \frac{1}{n!} \int dx_1 \cdots dx_n \, \phi_n(x; x_1, \ldots, x_n) |x_1, \ldots, x_n\rangle$$

and (C.54) we obtain the solution to (4.69) of the form

$$|\phi(x)\rangle = \exp \left( -\frac{1}{2\nu} \int_c^x dx \, a^\dagger(x) \right) |0\rangle,$$

where $c$ is an arbitrary constant. Hence, taking into account (4.67) we finally arrive at the desired linearization transformation for the Burgers equation such that

$$\phi[u|x] = \exp \left( -\frac{1}{2\nu} \int_c^x dx \, u \right). \tag{4.72}$$

As an immediate consequence of (4.72), we find that the inverse transformation

$$u = -2\nu \partial_x \phi / \phi \tag{4.73}$$

reduces solution of (4.68) to the solution of its linear part subject to the initial condition

$$\exp \left( -\frac{1}{2\nu} \int_c^x dx \, u_0 \right) \tag{4.74}$$

Evidently, (4.73) coincides with the celebrated Hopf-Cole transformation reducing solutions of the Burgers equation to solutions of the heat equation. It should also be noted that the formulae (4.73) and (4.74) obtained by means of the actual treatment hold true regardless of the square integrability of $u_0$. □

REMARK. The linearization technique described above can be treated as a generalization of the classical method of variation of constants to the case of nonlinear partial differential equations. Indeed, the actual treatment for linearization of nonlinear partial differential equations is the most natural development of the algorithm for linearization of ordinary differential equations introduced in previous section. On the other hand, any procedure for the reduction of a system of nonlinear ordinary differential equations to the solution of its linear part like that presented in previous section has to be equivalent to the method of variation of constants. Let us only recall the "interaction picture" within the Hilbert space approach discussed in section 1.2.

Finally, we note that the classical method of variation of constants seems to fail in the case of nonlinear partial differential equations. This observation can be illustrated by the example of the Burgers equation

$$\partial_t u = \nu \partial_x^2 u - u \partial_x u. \tag{4.75}$$

If we set in (4.75)

$$u = \exp(\nu t \partial_x^2) v(x,t),$$

we find

$$\partial_t v = -\sum_{i=0}^{\infty} \frac{(-2\nu t)^i}{i!} \partial_x^i v \partial_x^{i+1} v.$$

This nonlinear, infinite-order equation is more complicated than the original equation (4.75). We conclude that the standard method of variation of constants does not work in the case of the Burgers equation. ■

## 4.3 Bäcklund transformations

Our aim in this section is to extend the observations of section 4.2.2 to the case involving general Bäcklund transformations. This leads to a method of determining Bäcklund transformations for nonlinear partial differential equations of the evolution type [39]. Consider the following equation:

$$\partial_t u(x,t) = F(u, D^\alpha u), \qquad u(x,0) = u_0(x), \tag{4.76}$$

where $u$: $\mathbf{R}^s \times \mathbf{R} \to \mathbf{R}$, $F$ is analytic in $u$, $D^\alpha u$ and $u_0 \in L^2_{\mathbf{R}}(\mathbf{R}^s, d^s x)$. Let us write down the "Hamiltonian" corresponding to (4.76)

$$M = \int d^s x \, a^\dagger(x) F(a(x), D^\alpha a(x)).$$

As we have already seen, the transformation

$$u' = \phi[u|x], \tag{4.77}$$

where $\phi$ is analytic in $u$, leads to the following transformation law obeyed by the "Hamiltonian" $M$:

$$M' = \int d^s x \, a^\dagger(x) [\phi[a|x], M]. \tag{4.78}$$

Consider now the general case of the transformation (4.77) converting eq. (4.76) into

$$\partial_t u' = F'(u', D^\beta u'). \tag{4.79}$$

From (4.78), we have

$$[\phi[a|x], M] = F'(\phi[a|x], D^\beta \phi[a|x]). \tag{4.80}$$

Taking the Hermitian conjugate of (4.80) we obtain the following equation in Hilbert space:

$$M^{\dagger}|\phi(x)\rangle = F'(\phi[a^{\dagger}|x], D^{\beta}\phi[a^{\dagger}|x])|0\rangle, \tag{4.81}$$

where $|\phi(x)\rangle = \phi[a^{\dagger}|x]|0\rangle$.

Clearly, the vector $|\phi(x)\rangle$ satisfying (4.81) and the Bäcklund transformation (4.77) are related by

$$\phi[u|x] = \langle u|\phi(x)\rangle \exp\left(\frac{1}{2}\int d^s x\, u^2\right). \tag{4.82}$$

We have thus reduced the problem of finding Bäcklund transformations to the solution of the abstract equation in Hilbert space (4.81). The reader may notice that (4.81) is linear only in the case of the linearization transformation, when (4.79) is of the form (4.65) and (4.81) reduces to (4.66). Nevertheless, it turns out that there exist nontrivial cases when the Bäcklund transformations can be determined easily by solving (4.81). For an easy illustration of this observation we now rederive the Miura transformation.

EXAMPLE. Consider the modified Korteweg-de Vries equation

$$\partial_t u = -\partial_x^3 u + 6u^2\partial_x u \tag{4.83}$$

and the Korteweg-de Vries equation

$$\partial_t u' = -\partial_x^3 u' + 6u'\partial_x u'. \tag{4.84}$$

We look for the Bäcklund transformation $\phi$ converting (4.83) into (4.84), i.e.,

$$u' = \phi[u|x]. \tag{4.85}$$

The abstract equation (4.81) becomes

$$M^{\dagger}|\phi(x)\rangle = -\partial_x^3|\phi(x)\rangle + 6\phi[a^{\dagger}|x]\partial_x|\phi(x)\rangle, \tag{4.86}$$

where $M^{\dagger}$ is the conjugate of the "Hamiltonian" corresponding to (4.83) such that

$$M^{\dagger} = \int dx\,[-a^{\dagger\prime\prime\prime}(x) + 6a^{\dagger 2}(x)a^{\dagger\prime}(x)]a(x).$$

On writing (4.86) in the coordinate representation we get

$$\partial_{x_1}^3\phi_1(x;x_1) = -\partial_x^3\phi_1(x;x_1), \tag{4.87a}$$

$$\begin{aligned}(\partial_{x_1}^3 + \partial_{x_2}^3)\phi_2(x;x_1,x_2) &= -\partial_x^3\phi_2(x;x_1,x_2)\\ &+ 6\phi_1(x;x_2)\partial_x\phi_1(x;x_1) + 6\phi_1(x;x_1)\partial_x\phi_1(x;x_2),\end{aligned} \tag{4.87b}$$

$$\sum_{i=1}^{n+2}\partial_{x_i}^3\phi_{n+2}(x;x_1,\ldots,x_{n+2})$$
$$-12\sum_{i=1}^{n+2}\sum_{\substack{r,s\neq i\\ r>s}}\partial_{x_i}[\delta(x_i-x_r)\delta(x_i-x_s)\phi_n(x;x_1,\ldots,\check{x}_r,\ldots,\check{x}_s,\ldots,x_{n+2})]$$
$$=-\partial_x^3\phi_{n+2}(x;x_1,\ldots,x_{n+2})$$
$$+6\sum_{r=1}^{n+1}\frac{1}{r!}\sum_{i_1=1}^{n+2}\sum_{i_2\neq i_1}\cdots\sum_{i_r\neq i_{r-1}}\phi_{n+2-r}(x;x_1,\ldots,\check{x}_{i_1},\ldots,\check{x}_{i_r},\ldots,x_{n+2})$$
$$\times\,\partial_x\phi_r(x;x_{i_1},\ldots,x_{i_r}),\qquad n=1,2,\ldots,\infty, \tag{4.87c}$$

where $\phi_n(x;x_1,\ldots,x_n)=\langle x_1,\ldots,x_n|\phi(x)\rangle$ and the reversed hat over $x_r$, $x_s$ and $x_{i_1}$, $x_{i_r}$ denotes that these variables should be omitted from the set $\{x_1,\ldots,x_{n+2}\}$.

Performing the Fourier transformation we obtain from (4.87) the following equations:

$$(k^3+k_1^3)\tilde{\phi}_1(k;k_1)=0, \tag{4.88a}$$
$$(k^3+k_1^3+k_2^3)\tilde{\phi}_2(k;k_1,k_2)=-6k\int dk'\,\tilde{\phi}_1(k-k';k_1)\tilde{\phi}_1(k';k_2), \tag{4.88b}$$
$$\left(k^3+\sum_{i=1}^{n+2}k_i^3\right)\tilde{\phi}_{n+2}(k;k_1,\ldots,k_{n+2})$$
$$+12\sum_{i=1}^{n+2}\sum_{\substack{r,s\neq i\\ r>s}}k_i\tilde{\phi}_n(k;k_1,\ldots,\check{k}_r,\ldots,\check{k}_s,\ldots,k_{n+2})|_{k_i\to k_i+k_r+k_s}$$
$$=-6\sum_{r=1}^{n+1}\frac{1}{r!}\sum_{i_1=1}^{n+2}\sum_{i_2\neq i_1}\cdots\sum_{i_r\neq i_{r-1}}\int dk'$$
$$\times\,\tilde{\phi}_{n+2-r}(k-k';k_1,\ldots,\check{k}_{i_1},\ldots,\check{k}_{i_r},\ldots,k_{n+2})$$
$$\times\,k'\tilde{\phi}_r(k';k_{i_1},\ldots,k_{i_r}),\qquad n=1,2,\ldots,\infty. \tag{4.88c}$$

Using the identities

$$(k^3+k_1^3)\delta(k+k_1)=0,$$
$$\sum_{i=0}^{n}k_i^3\delta\left(\sum_{i=0}^{n}k_i\right)=3\sum_{q>r>s}k_qk_rk_s\delta\left(\sum_{i=0}^{n}k_i\right),$$

where $q$, $r$, $s\in\{0,1,\ldots,n\}$, $n\geq 2$, and we put $k_0=k$, the following solution of (4.88) can be found easily:

$$\tilde{\phi}_1=-ik\delta(k+k_1),\qquad \tilde{\phi}_2=2\delta(k+k_1+k_2),\qquad \tilde{\phi}_n=0,\quad n\geq 3. \tag{4.89}$$

On taking the inverse Fourier transformation of (4.89) and making use of

$$|\phi(x)\rangle=\sum_{n=1}^{\infty}\frac{1}{n!}\int dx_1\cdots dx_n\,\phi_n(x;x_1,\ldots,x_n)|x_1,\ldots,x_n\rangle,$$

we arrive at the solution of (4.86) such that

$$|\phi(x)\rangle = |xx\rangle + \partial_x|x\rangle.$$

Using (4.82) and (C.63) we finally obtain the desired Bäcklund transformation (4.85) of the form

$$u' = u^2 + \partial_x u. \tag{4.90}$$

We have thus rediscovered the famous Miura transformation which relates the Korteweg-de Vries equation to the modified Kortweg-de Vries equation. Actually, Miura showed that

$$\partial_t u' + \partial_x^3 u' - 6u'\partial_x u' = (2u + \partial_x)(\partial_t u + \partial_x^3 u - 6u^2\partial_x u). \tag{4.91}$$

Therefore, if $u$ is the solution of (4.83) then $u'$ given by (4.90) is the solution of (4.84). An easy inspection now shows that (4.91) can be written in the following form:

$$\mathrm{D}_{u'}F(u) = F'(u'), \tag{4.92}$$

where $\mathrm{D}_f g \equiv f'[g]$ designates the Gateaux derivative of $f(u)$ (see appendix E); $F(u)$, $F'(u)$ are the vector fields corresponding to the modified Korteweg-de Vries equation and Korteweg-de Vries equation, respectively, and we have used the shorthand notation $G(u) \equiv G(u, D^\gamma u)$.

The reader can easily check that (4.92) is nothing but the Bargmann realization (see appendix C) for the abstract equation (4.81). It thus appears that the standard approach corresponding to eq. (4.92) is included by the actual canonical Hilbert space formalism as a special case of the Bargmann representation for the abstract equation (4.81). □

REMARK. We note that the introduced method for finding Bäcklund transformations is not restricted to completely integrable equations. In fact, the method, when applied for determining linearization transformations, has been already shown to work in the case of the Burgers equation which is not completely integrable.

We point out that regardless of the form of eqs. (4.88b) and (4.88c) we have derived the solution (4.89) from (4.88) in purely algebraic manner. Indeed, in order to find the solution of the closed system (4.88a) and (4.88b) which coincides with (4.89) we need not have solved any integral equation. It should also be noted that it is enough to check the condition $\tilde{\phi}_n = 0,\ n \geq 3$, only for $\tilde{\phi}_3$ and $\tilde{\phi}_4$. ■

## 4.4 Feigenbaum-Cvitanovic renormalization equations

In this section we apply the observations presented in section 3.3 to study the Feigenbaum-Cvitanovic renormalization equations (universal equations) [14]. Consider the universal equation such that [40]:

$$f(\beta x) = \beta f(f(x)), \tag{4.93}$$

where $f$ is analytic in $x$ and $\beta$ is a scaling parameter.

Evidently, (4.93) amounts a particular case of (3.26). Taking into account (C.2a) and (C.14) we find that the Hilbert space equation (3.29) corresponding to (4.93) can be written in the form

$$M^\dagger|f\rangle = \beta^{N-1}|f\rangle, \tag{4.94}$$

where $|f\rangle = f(a^\dagger)|0\rangle$ and the operator $\beta^N$ is given by (C.15). Clearly, the solution to (4.94) is linked to the solution of (4.93) by

$$f(x) = \langle x|f\rangle e^{\frac{1}{2}x^2},$$

where $|x\rangle$ is a normalized coherent state.

It thus appears that the universal equation can be cast into the abstract equation in Hilbert space (4.94). On using the identities $\phi(x) \equiv \beta f(x)$ and $\psi(x) \equiv f(\beta x)$ we arrive at the following form of relations (3.30) corresponding to the occupation number representation for the abstract equation (4.94):

$$f(f_0) = \beta^{-1} f_0, \tag{4.95a}$$

$$\sum_{i=1}^{k} P_{ik}[f_0', \ldots, f_0^{(k-i+1)}] f^{(i)}(f_0) = \beta^{k-1} f_0^{(k)}, \qquad k = 1, 2, \ldots, \infty, \tag{4.95b}$$

where $P_{ik}$ are given by (3.31).

EXAMPLE. Consider the ansatz (3.38), where $\phi = f$. On setting $\tilde{g}(N) \equiv \nu$, $\tilde{h}(N) \equiv 1$ we obtain at once

$$f(a^\dagger) = \frac{a^\dagger}{1 + \nu a^\dagger}. \tag{4.96}$$

Inserting (4.96) into (4.93) we arrive at the following solution to the universal equation (4.93), where $\beta = 2$ [41]:

$$f(x) = \frac{x}{1 - \gamma x}, \tag{4.97}$$

where $\gamma$ is a constant. □

In opposite to the limited role of equations (3.36) for determining the linearization transformations for nonlinear difference equations, the relations (4.95) corresponding to the occupation number representation for the abstract equation (4.94) appear to be effective in finding solutions of the universal equation (4.93). We now illustrate this observation by the following example.

EXAMPLE. Consider equations (4.95). We first discuss the case $f_0 \neq 0$. Calculating (4.95b) for $k = 1, 2$, letting $f_0'' \neq 0$ and utilizing (4.95a) we obtain the following equation:

$$f'(f_0) = \frac{f_0}{f(f_0)}.$$

The solution of this trivial equation is

$$f(x) = \sqrt{x^2 + \gamma}, \tag{4.98}$$

where $\gamma$ is an integration constant.

It follows immediately from (4.95a) that (4.98) is the solution of the universal equation (4.93), where $\beta = 1/\sqrt{2}$. Moreover, it is easy to verify that the function (4.98) is a member of the family of solutions to (4.93) such that

$$f_\alpha(x) = (x^\alpha + \gamma)^{1/\alpha},$$

where the corresponding scaling parameters are $\beta_\alpha = 2^{-1/\alpha}$

Now we pass to the case $f_0 = 0$. On assuming $f_0 = 0$ and making use of (4.95b) the following solution of (4.93) can be obtained easily:

$$f_0' = 1, \qquad \beta = 2, \qquad f_0^{(i)} = \frac{i!}{2^{i-1}} f_0''^{i-1}, \quad i = 2, 3, \ldots,$$

where $f_0'' \neq 0$. On putting $f_0''/2 = \gamma$ we find that the corresponding solution of the universal equation (4.93) is

$$f(x) = \sum_{i=1}^{\infty} \gamma^{i-1} x^i = \frac{x}{1 - \gamma x}.$$

We have thus obtained the solution (4.97). Using (4.95b) the author established the existence of the solution to (4.93) of the form

$$f_0 = 0, \qquad f_0' = 1, \qquad f_0'' = 0, \qquad \beta = \sqrt{2}, \qquad f_0^{(2i)} = 0,$$
$$f_0^{(2i+1)} = c_i f_0'''^i, \qquad i = 2, 3, \ldots,$$

where $f_0''' \neq 0$. A few first terms in the power series expansion for the corresponding universal function $f(x)$ satisfying (4.93), where $\beta = \sqrt{2}$, are given by

$$f(x) = x + \frac{1}{3!}x^3 + \frac{10}{2 \cdot 5!}x^5 + \frac{350}{6 \cdot 7!}x^7 + \frac{17150}{14 \cdot 9!}x^9 + O(x^{11}),$$

where we set $f_0''' = 1$. □

## 4.5 Chaos

Our aim in this section is to formulate a hypothesis concerning "quantal" Hilbert space criterion for the classical dynamical systems to exhibit chaotic behaviour [42]. Consider the following nonlinear dynamical system (complex or real):

$$\frac{d\mathbf{z}}{dt} = \mathbf{F}(\mathbf{z}), \qquad \mathbf{z}(0) = \mathbf{z}_0, \tag{4.99}$$

where $\mathbf{F}$: $\mathbf{C}^k \to \mathbf{C}^k$ is analytic in $\mathbf{z}$.

We define the operator $\rho$ as follows (see (C.29)):

$$\rho = \int_{\mathbf{R}^{2k}} d\mu(\mathbf{w})\, e^{-|\mathbf{w}-\mathbf{z}_0|^2} |\mathbf{w}\rangle\langle\mathbf{w}|, \tag{4.100}$$

where $|\mathbf{w}\rangle$, $\mathbf{w} \in \mathbf{C}^k$, are the normalized coherent states.

The following properties of the operator $\rho$ can be established easily:

$$\rho^\dagger = \rho, \tag{4.101a}$$

$$\rho \geq 0, \tag{4.101b}$$

$$\mathrm{Tr}\rho = 1. \tag{4.101c}$$

We note that the simplest way of derivation of (4.101c) is to apply (C.40). Furthermore, eqs. (4.100), (C.29), (C.38) and (1.34) taken together yield

$$\langle \mathbf{a}(t)\rangle = \mathrm{Tr}(\rho\mathbf{a}(t)) = \mathbf{z}(t), \tag{4.102}$$

where $\mathbf{a}(t)$ are the time-dependent Bose annihilation operators discussed in section 1.2, $\langle \mathbf{a}(t)\rangle$ is the expectation value of $\mathbf{a}(t)$ (see appendix B) and $\mathbf{z}(t)$ satisfies (4.99).

Recall that we deal within the "quantum-mechanical" Hilbert space formulation for nonlinear dynamical systems described in chapter 1, with a "quantization" scheme $\mathbf{z} \to \mathbf{a}$. We conclude from the form of (4.101) and (4.102) that $\rho$ is the "density matrix" (see appendix B) in the "quantal" approach. Now, in analogy to quantum mechanics, we can introduce the "entropy" such that

$$S = -\mathrm{Tr}(\rho \ln \rho). \tag{4.103}$$

To calculate the value of this expression, we first need to compute $\mathrm{Tr}\rho^n$ Using (4.100), (C.29) and (C.37) we get

$$\mathrm{Tr}\rho^n = \frac{1}{2^n - 1}, \qquad n \geq 1.$$

Hence, taking into account (4.103) we find, after some calculation, that

$$S = -\sum_{i=0}^{\infty} \frac{1}{2^i} \ln \frac{1}{2^i} = 2 \ln 2.$$

The obtained value of the "entropy" can be interpreted as an averaged amount of the information needed to specify the coherent state $|\mathbf{z}_0\rangle$ in a Hilbert space of states. Indeed, an arbitrary vector in a Fock space can be represented by an infinite sequence of $i$-vectors of 1's and 0's. On the other hand, the number of different $i$-vectors is $2^i$.

We now return to (4.100). Proceeding analogously as in quantum mechanics, we define the time-dependent "density matrix" as

$$\rho(t) = \int_{\mathbf{R}^{2k}} d\mu(\mathbf{w})\, e^{-|\mathbf{w}-\mathbf{z}_0|^2} e^{tM}|\mathbf{w}\rangle\langle\mathbf{w}|e^{tM^\dagger}, \tag{4.104}$$

where $M = \mathbf{a}^\dagger \cdot \mathbf{F}(\mathbf{a})$ is the "Hamiltonian" corresponding to (4.99).

It is easy to verify that $\rho(t)$ is Hermitian and nonnegative at any time. As in (4.103) we can introduce the time-dependent "entropy" by

$$S(t) = -\mathrm{Tr}[\rho(t)\ln\rho(t)].$$

We are now ready to formulate our hypothesis. Namely, we claim that if the following conditions are valid:

$$\underset{t_*>0}{\exists}\ \underset{t>t_*}{\forall}\ |\mathbf{z}(\mathbf{z}_0,t)|^2 < C, \tag{4.105a}$$

where $\mathbf{z}(\mathbf{z}_0,t)$ is the solution to (4.99); $C > 0$ is a constant, and

$$\underset{t>t_*}{\forall}\ \frac{dS(t)}{dt} > 0, \tag{4.105b}$$

then the system (4.99) shows chaotic behaviour.

REMARK 1. We note that the condition (4.105a) ensures that $\mathrm{Tr}\rho(t) < \infty$ for $t > t_*$. In fact, using (4.104) and the formula

$$\langle\mathbf{z}_0,t|\mathbf{z}_0,t\rangle = \exp(|\mathbf{z}(\mathbf{z}_0,t)|^2 - |\mathbf{z}_0|^2),$$

which can be easily obtained from (1.2a) (see also (1.5)), we find

$$\mathrm{Tr}\rho(t) = \int_{\mathbf{R}^{2k}} d\mu(\mathbf{w})\, e^{-|\mathbf{w}-\mathbf{z}_0|^2} e^{|\mathbf{z}(\mathbf{w},t)|^2 - |\mathbf{w}|^2}. \qquad \blacksquare$$

REMARK 2. Recall that in quantum mechanics we deal with unitary evolution operators which leads to $S(t) = S = \text{const}$. Notice that the evolution operator $e^{tM}$ is unitary only in the case of the linear system (4.99) with a skew-Hermitian matrix. ■

REMARK 3. It is a natural question to ask whether the "quantal" Hilbert space approach recognizes the divergence of orbits in the phase space of the system (4.99). An affirmative answer is provided by the following relation which is an immediate consequence of (1.2a) and (C.28b):

$$\frac{|\langle\mathbf{z}_0,t|\mathbf{z}_0+\delta\mathbf{z}_0,t\rangle|^2}{\langle\mathbf{z}_0,t|\mathbf{z}_0,t\rangle\langle\mathbf{z}_0+\delta\mathbf{z}_0,t|\mathbf{z}_0+\delta\mathbf{z}_0,t\rangle} = \exp(-|\mathbf{z}(\mathbf{z}_0+\delta\mathbf{z}_0,t) - \mathbf{z}(\mathbf{z}_0,t)|^2). \tag{4.106}$$

This formula shows that the growth of the distance between the two trajectories of the system (4.99) starting from $\mathbf{z}_0$ and $\mathbf{z}_0+\delta\mathbf{z}_0$ implies the decay of the corresponding transition probability from the left-hand side of (4.106). ■

APPENDIX A

# HILBERT SPACES

As we promised in the preface, we now summarize the basic facts about the formalism of Hilbert spaces within the *Dirac bra-ket notation* [43] which is used throughout this book. Let $\mathcal{H}$ be a Hilbert space. We assume that the inner product $\langle\cdot|\cdot\rangle\colon \mathcal{H}\times\mathcal{H}\to \mathbf{C}$ is linear in the second argument. By virtue of the Riesz theorem, an arbitrary linear, continuous functional on $\mathcal{H}$ is of the form

$$\phi \to \langle\psi|\phi\rangle, \tag{A.1}$$

where $\psi\in\mathcal{H}$.

Conversely, an arbitrary vector $\psi\in\mathcal{H}$, defines the linear continuous functional (A.1). We note that the vector $\psi$ is uniquely determined by (A.1). Applying the Dirac notation we use the symbol $\langle\psi|$ for the functional (A.1) belonging to the Hilbert space $\mathcal{H}^*$ of linear, continuous functionals on $\mathcal{H}$, and $|\psi\rangle$ for the vector $\psi\in\mathcal{H}$. The vectors $\langle\psi|$ and $|\psi\rangle$ are called *bra-vectors* and *ket-vectors*, respectively. Clearly, $|\psi\rangle\to\langle\psi|$ is the one-one antilinear map between $\mathcal{H}$ and $\mathcal{H}^*$. Following the Dirac's convention we identify $\mathcal{H}$ with $\mathcal{H}^*$ and we treat the inner product $\langle\phi|\psi\rangle$ as the product of $\langle\phi|$ and $|\psi\rangle$ obtained formally by the graphical joining of $\langle\phi|$ with $|\psi\rangle$.

EXAMPLE. Consider the Hilbert space $l^2$ specified by the inner product

$$\langle\phi|\psi\rangle = \sum_{i=1}^{\infty}\phi_i^*\psi_i.$$

The ket-vector $|\psi\rangle$ and bra-vector $\langle\phi|$ correspond to the infinite column-vector

$$\begin{bmatrix}\psi_1\\ \psi_2\\ \vdots\end{bmatrix},$$

and the infinite row-vector $[\phi_1^*,\phi_2^*,\ldots]$, respectively. □

We now discuss the action of operators on the vectors in a Hilbert space within the Dirac notation. As the reader may have already noticed, we can give up the notation $\langle\phi|L\psi\rangle$, where $L$: $\mathcal{H} \to \mathcal{H}$ is a linear operator. Instead, we can write $\langle\phi|L|\psi\rangle$. Therefore, the Hermitian conjugate $L^\dagger$ of a linear operator $L$ is defined by

$$\langle\phi|L^\dagger|\psi\rangle = (\langle\psi|L|\phi\rangle)^* \quad \text{for arbitrary} \quad |\phi\rangle, |\psi\rangle \in \mathcal{H}, \tag{A.2}$$

where the asterisk designates the complex conjugation.

Analogously, the unitary operator $U$ satisfies

$$\langle\phi|U^\dagger U|\psi\rangle = \langle\phi|\psi\rangle.$$

Taking into account (A.2) we find

$$L|\phi\rangle = |\psi\rangle \quad \text{if and only if} \quad \langle\psi| = \langle\phi|L^\dagger. \tag{A.3}$$

In particular, if $|\lambda\rangle$ is an eigenvector of the operator $L$ corresponding to the eigenvalue $\lambda$, then

$$L|\lambda\rangle = \lambda|\lambda\rangle \quad \text{if and only if} \quad \lambda^*\langle\lambda| = \langle\lambda|L^\dagger.$$

It should also be noted that (A.3) can be formally written as

$$(L|\phi\rangle)^\dagger = \langle\phi|L^\dagger,$$

whence

$$(\alpha|\phi\rangle)^\dagger = \alpha^*\langle\phi|. \tag{A.4}$$

We now proceed to study projections in the Dirac notation. We introduce the linear operator $|\phi_1\rangle\langle\phi_2|$ such that

$$(|\phi_1\rangle\langle\phi_2|)|\psi\rangle = \langle\phi_2|\psi\rangle|\phi_1\rangle, \tag{A.5a}$$

$$\langle\psi|(|\phi_1\rangle\langle\phi_2|) = \langle\psi|\phi_1\rangle\langle\phi_2|. \tag{A.5b}$$

The operator $|\phi_1\rangle\langle\phi_2|$ is usually known as the diadic or "outer" product of $|\phi_1\rangle$ and $\langle\phi_2|$. Note that the vectors from the right-hand side of (A.5a) and (A.5b) are formally obtained by "joining" $|\phi_1\rangle\langle\phi_2|$ to $|\psi\rangle$ and $\langle\psi|$, respectively.

EXAMPLE. Referring to the example above, we find that the operator $|\phi\rangle\langle\psi|$ corresponds to the infinite matrix with elements $[|\phi\rangle\langle\psi|]_{ij} = \phi_i\psi_j^*$, $i, j = 1, 2, \ldots, \infty$. □

The following properties of the operators $|\cdot\rangle\langle\cdot|$ can be established easily:

$$(|\phi\rangle\langle\psi|)^\dagger = |\psi\rangle\langle\phi|, \tag{A.6}$$

$$(|\phi_1\rangle\langle\psi_1|)(|\phi_2\rangle\langle\psi_2|) = |\phi_1\rangle\langle\psi_1|\phi_2\rangle\langle\psi_2| = \langle\psi_1|\phi_2\rangle|\phi_1\rangle\langle\psi_2|. \tag{A.7}$$

Now, let $|\phi\rangle$ be a normalized vector. It follows immediately from (A.6) and (A.7) that the operator $|\phi\rangle\langle\phi|$ is Hermitian and idempotent, i.e. it is a projection on the vector $|\phi\rangle$. Therefore, the resolution of the identity in the Dirac notation takes the form

$$\sum_i |i\rangle\langle i| = I, \tag{A.8}$$

where the vectors $|i\rangle$ form a complete orthonormal set (orthonormal basis) in a separable Hilbert space $\mathcal{H}$, that is $\langle i|j\rangle = \delta_{ij}$ and $\langle\phi|i\rangle = 0$ for arbitrary $i$, implies $|\phi\rangle = 0$.

An advantage of the Dirac notation is that the symbols like (A.8) allow to perform some calculations almost automatically. For example, the expansion of the vector $|\phi\rangle$ in the basis $|i\rangle$ is simply

$$|\phi\rangle = \sum_i \langle i|\phi\rangle|i\rangle.$$

Furthermore, if we want to calculate the Fourier coefficient of $|\phi\rangle$ with respect to the basis $|\alpha\rangle$, we immediately obtain from (A.8) that

$$\langle\alpha|\phi\rangle = \sum_i \langle\alpha|i\rangle\langle i|\phi\rangle.$$

On extending the usual Hilbert space to the space of distributions (the resulting Hilbert space is called the rigged Hilbert space) we can introduce the basis $|x\rangle$ (see appendix D) marked with a continuous real parameter $x$, such that the norm of $|x\rangle$ is infinite. The basis vectors are normalized by the Dirac delta function, i.e. $\langle x|x'\rangle = \delta(x - x')$. The corresponding resolution of the identity is given by

$$\int dx\, |x\rangle\langle x| = I.$$

On the other hand, passing to the complex domain we can deal with the nonorthogonal basis $|z\rangle$ marked with a complex number $z$, satisfying $\langle z|z\rangle = 1$, $\langle z|z'\rangle \neq 0$ for $z \neq z'$. The nonorthogonal resolution of the identity related to such basis

$$\int d\mu(z)\, |z\rangle\langle z| = I,$$

implies that $|z\rangle$'s form the overcomplete set.

Finally, we discuss the tensor product of Hilbert spaces in the Dirac notation. Let $\mathcal{H}_1 \otimes \mathcal{H}_2$ be the tensor product of Hilbert spaces $(\mathcal{H}_1, \langle\cdot|\cdot\rangle_{\mathcal{H}_1})$ and $(\mathcal{H}_2, \langle\cdot|\cdot\rangle_{\mathcal{H}_2})$. Recall that the inner product in $\mathcal{H}_2 \otimes \mathcal{H}_2$ is given by

$$\langle\phi_1 \otimes \psi_1|\phi_2 \otimes \psi_2\rangle = \langle\phi_1|\phi_2\rangle_{\mathcal{H}_1}\langle\psi_1|\psi_2\rangle_{\mathcal{H}_2}. \tag{A.9}$$

Taking into account the form of (A.9) it follows that the tensor product of vectors $\phi$ and $\psi$ is written in the Dirac notation as

$$|\phi \otimes \psi\rangle = |\phi\rangle|\psi\rangle,$$

where $|\phi\rangle \in \mathcal{H}_1$ and $|\psi\rangle \in \mathcal{H}_2$.

We note that formally (see (A.4)):

$$(|\phi\rangle|\psi\rangle)^\dagger = \langle\psi|\langle\phi|.$$

It should also be noted that we can define the action of the linear operators $L_1$: $\mathcal{H}_1 \to \mathcal{H}_1$ and $L_2$: $\mathcal{H}_2 \to \mathcal{H}_2$ on the tensor product of $|\phi\rangle \in \mathcal{H}_1$ and $|\psi\rangle \in \mathcal{H}_2$ by

$$L_1(|\phi\rangle|\psi\rangle) = (L_1|\phi\rangle)|\psi\rangle,$$
$$L_2(|\phi\rangle|\psi\rangle) = |\phi\rangle(L_2|\psi\rangle).$$

The operators $L_1$ and $L_2$ are usually designated by $L_1 \otimes I$ and $I \otimes L_2$, respectively.

EXAMPLE. As we have seen in the two previous examples, the bra-vector $|\phi\rangle$ is represented in the Hilbert space $l^2$ by an infinite column-vector. The reader can easily check that the vector $|\phi\rangle|\psi\rangle$ can be identified with the direct (Kronecker) product of infinite colum-vectors corresponding to $|\phi\rangle$ and $|\psi\rangle$, respectively. □

Appendix B

# QUANTUM MECHANICS

We now briefly introduce the basic ideas and definitions of quantum mechanics. We begin with the postulates of quantum statics which specializes to the study of a physical system at a particular instant of time.

(i) The states of a physical system are unit vectors of a complex, separable, infinite-dimensional Hilbert space $\mathcal{H}$. More precisely, they are rays (one-dimensional subspaces) in such a Hilbert space.

(ii) If a physical system is in the state $|\phi\rangle$, then the probability that it will be actually found in the state $|\psi\rangle$ when an experiment is performed to decide whether it is in the state $|\psi\rangle$ or not, is $|\langle\phi|\psi\rangle|^2$. This probability is known as a *transition probability* between states $|\phi\rangle$ and $|\psi\rangle$.

(iii) A physical quantity or observable is described by a linear Hermitian operator. After a measurement of an observable the state of a physical system is an eigenvector of the corresponding Hermitian operator. The possible results of an experiment are eigenvalues of the Hermitian operator related to the measured physical quantity. In a given state of the system, an observable has only a probability distribution and in general no sharply defined value. If the system is in the state $|\phi\rangle$, then the probability that the measurement of the observable $L$ gives the result from a Borel set $\Delta$ is

$$p_{L,\phi}(\Delta) = \langle\phi|E_L(\Delta)|\phi\rangle \qquad \text{(B.1a)}$$

$$= \int_\Delta |\phi(\lambda)|^2\, d\lambda, \qquad \text{(B.1b)}$$

where $\phi(\lambda) = \langle\lambda|\phi\rangle$ and $|\lambda\rangle$ is an eigenvector of $L$ corresponding to the continuous eigenvalue $\lambda$; the map

$$\Delta \to E_L(\Delta) = \int_\Delta d\lambda\, |\lambda\rangle\langle\lambda|$$

is a projection-valued measure giving the spectral decomposition of the Hermitian operator $L$ such that

$$L = \int \lambda|\lambda\rangle\langle\lambda|\, d\lambda. \qquad \text{(B.2)}$$

In the case of the purely discrete simple spectrum of $L$ say, $\lambda_i$, $i = 1, 2, \ldots$, the probability that a physical quantity described by $L$ is $\lambda_k$, is given by

$$p_{L,\phi}(\lambda_k) = |\phi_k|^2,$$

where $\phi_k = \langle k|\phi\rangle$, $L|i\rangle = \lambda_i|i\rangle$, $i = 1, 2, \ldots$, and $\langle i|j\rangle = \delta_{ij}$.

REMARK. It is customary in quantum mechanics to say interchangeably of observables and the corresponding Hermitian operators. A procedure for the construction of observables describing the classical quantities is called *quantization*. ■

We note that the map $\Delta \to p_{L,\phi}(\Delta)$ given by (B.1) is a probability measure. By analogy with classical probability theory it is called the distribution of the observable $L$ in the state $|\phi\rangle$. Taking into account (B.2) we find that the corresponding *expectation value* of $L$ in the (normalized) state $|\phi\rangle$ is

$$\langle L\rangle = \langle\phi|L|\phi\rangle.$$

Using this we arrive at the following formula on the *variance* of $L$ in the state $|\phi\rangle$:

$$\Delta L^2 = \langle(L - \langle L\rangle)^2\rangle = \langle L^2\rangle - \langle L\rangle^2. \tag{B.3}$$

Clearly, the *standard deviation* $\Delta L$ giving a measure of the dispersion of $L$ around its expectation value is obtained from (B.3) by $\Delta L = \sqrt{\Delta L^2}$. The standard deviations of arbitrary observables $A$ and $B$ satisfy the following *Heisenberg uncertainty principle*:

$$\Delta A \Delta B \geq \frac{1}{2}|\langle[A, B]\rangle|. \tag{B.4}$$

This inequality shows that if the observables $A$ and $B$ do not commute with one another, then they cannot be measured simultaneously with an arbitrary accuracy. The famous example of (B.4) are the *Heisenberg relations*

$$\Delta\hat{q}_i \Delta\hat{p}_i \geq \frac{\hbar}{2}, \qquad i = 1, 2, 3, \tag{B.5}$$

where $\hat{q}_i$ and $\hat{p}_i$ are position and momentum observables, respectively (see appendix D) and $\hbar = 1.055 \times 10^{-34}$ J s is Planck's constant devided by $2\pi$. The uncertainty relations characterize the range of the applicability of the classical manner of describing natural phenomena.

We now discuss the dynamical properties of quantum-mechanical systems. It is postulated that the motion of the states is described by the differential equation

$$i\frac{d}{dt}|\phi, t\rangle = H(t)|\phi, t\rangle, \qquad |\phi, 0\rangle = |\phi_0\rangle, \tag{B.6}$$

where $H(t)$ is the Hermitian operator corresponding to the energy of the system and we set $\hbar = 1$.

We call (B.6) the abstract *Schrödinger equation*. The operator $H(t)$ is called the *Hamiltonian* of the considered system. On introducing the unitary *evolution operator* $U(t)$ by

$$|\phi_0, t\rangle = U(t)|\phi_0\rangle, \tag{B.7}$$

where $|\phi_0, t\rangle$ is the solution to (B.6),

and inserting (B.7) into (B.6) we find that the Schrödinger equation (B.6) is equivalent to the following operator evolution equation:

$$i\frac{dU}{dt} = H(t)U, \qquad U(0) = I. \tag{B.8}$$

The formal solution of (B.8) given by the Neuman series solution of the integral equation equivalent to (B.8) such that

$$U(t) = I - i\int_0^t H(\tau)U(\tau)\,d\tau,$$

can be written as

$$U(t) = I + \sum_{n=1}^{\infty}(-i)^n \int_0^t d\tau_n \int_0^{\tau_n} d\tau_{n-1} \cdots \int_0^{\tau_2} d\tau_1\, H(\tau_n)\cdots H(\tau_1). \tag{B.9a}$$

The solution (B.9a) is often written in the following equivalent form:

$$U(t) = I + \sum_{n=1}^{\infty}(-i)^n \int_0^t d\tau_n \int_0^t d\tau_{n-1} \cdots \int_0^t d\tau_1\, H(\tau_n)\theta_{nn-1}H(\tau_{n-1})\cdots\theta_{21}H(\tau_1), \tag{B.9b}$$

where

$$\theta_{kl} = \theta(\tau_k - \tau_l) = \begin{cases} 1, & \tau_k \geq \tau_l, \\ 0, & \tau_k < \tau_l. \end{cases}$$

As is readily verified, if the following condition is satisfied:

$$[H(t), H(t')] = 0 \quad \text{for arbitrary} \quad t,\, t',$$

then the solution (B.9) becomes

$$U(t) = \exp\left(-i\int_0^t d\tau\, H(\tau)\right)$$

In particular case when the Hamiltonian is independent of time the evolution operator reduces to the usual exponential function of the operator $H$

$$U(t) = e^{-itH}. \tag{B.10}$$

We note that in the case of the basis $|E\rangle$ generated by the eigenvectors of the Hamiltonian $H$, the evolution operator (B.10) reduces to the phase factor $e^{iEt}$

The question naturally arises as to whether the evolution operator admits an exponential representation

$$U(t) = \exp[\Phi(t)], \tag{B.11}$$

where the Hermitian operator $-i\Phi(t)$ is called the *phase operator*,

in the case of time-dependent Hamiltonian $H(t)$. It was Magnus (see [44] and references therein) who first examined this problem. Namely, he found an iterative procedure which allows to develop the operator $\Phi(t)$ into terms of various orders in $H(t)$. Therefore, the expansion

$$\Phi(t) = \sum_{n=1}^{\infty} \Phi_n(t)$$

is usually known as the *Magnus expansion*. The explicit form of $\Phi_n$ for arbitrary $n$ was derived in ref. [45]. The canonical representation of $\Phi(t)$ introduced therein is given by

$$\Phi(t) = \sum_{n=1}^{\infty} \frac{(-i)^n}{n!} \int_0^t d\tau_n \cdots \int_0^t d\tau_1 \, (-1)^{n-\Theta_n-1}\Theta_n!(n-\Theta_n-1)! H(\tau_n)\cdots H(\tau_1),$$

where $\Theta_1 = 0$ and $\Theta_n = \Theta_n(\tau_n, \ldots, \tau_1) = \sum_{i=2}^{n} \theta_{ii-1}, \ n \geq 2.$

A few first terms in the Magnus expansion for the operator $\Phi(t)$ are [44,45]:

$$\begin{aligned}\Phi(t) = &-i\int_0^t d\tau \, H(\tau) - \frac{1}{2}\int_0^t d\tau_2 \int_0^{\tau_2} d\tau_1 \, [H(\tau_2), H(\tau_1)] \\ &+ \frac{i}{6}\int_0^t d\tau_3 \int_0^{\tau_3} d\tau_2 \int_0^{\tau_2} d\tau_1 \, \{[H(\tau_3),[H(\tau_2),H(\tau_1)]] + [H(\tau_1),[H(\tau_2),H(\tau_3)]]\} \\ &+ \cdots. \end{aligned} \tag{B.12}$$

The way of looking at dynamics of quantum-mechanical systems based on the assumption that the states change in time according to the Schrödinger equation and observables do not depend on time is referred to as *Schrödinger picture*. We now discuss an alternative language to describe the quantum dynamics which is called the *Heisenberg picture*. Suppose that the system is in the state $|\phi_0\rangle$. The probablity distribution of an observable $L$ at time $t$ is

$$p^{\mathrm{S}}_{L,\phi_0}(\Delta, t) = \langle \phi_0, t | E_L(\Delta) | \phi_0, t \rangle, \tag{B.13}$$

where $|\phi_0, t\rangle$ is the solution of the Schrödinger equation (B.6).

On using (B.7) we find that (B.13) can be written as

$$p^{\mathrm{H}}_{L,\phi_0}(\Delta, t) = \langle \phi_0 | U(t)^{-1} E_L(\Delta) U(t) | \phi_0 \rangle.$$

If we define

$$L(t) = U(t)^{-1} L U(t), \tag{B.14}$$

then $\Delta \to E_L^t(\Delta) = U(t)^{-1} E_L(\Delta) U(t)$ is the spectral measure of $L(t)$. Clearly, the probability distribution $p^{\mathrm{H}}_{L,\phi_0}$ is entirely described in terms of $E_L^t(\Delta)$ and $|\phi_0\rangle$. We may thus shift the time dependence to observables; the states remain unchanged. This way of looking at quantum dynamics is called the Heisenberg picture. On taking the (formal) time derivative of (B.14) and making use of (B.8) we arrive at the *Heisenberg equation of motion* satisfied by the time-dependent observable $L(t)$

$$i\frac{dL(t)}{dt} = [L(t), H(t)_{\mathrm{H}}], \qquad L(0) = L, \tag{B.15}$$

where

$$H(t)_{\mathrm{H}} = U(t)^{-1} H(t) U(t)$$

is the Hamiltonian in Heisenberg picture.

On using (B.11) and the operator identity

$$e^A B e^{-A} = B + \sum_{n=1}^{\infty} \frac{1}{n!} \underbrace{[A, \ldots, [A, B] \ldots]}_{n-\text{times}}, \tag{B.16}$$

we obtain the formal solution of (B.15) such that

$$L(t) = L + \sum_{n=1}^{\infty} \frac{(-1)^n}{n!} [\Phi(t), \ldots, [\Phi(t), L] \ldots].$$

If the Hamiltonian does not depend on time, then eqs. (B.10) and (B.11) readily yield $\Phi(t) = -itH$ and we get the power series expansion for $L(t)$ of the form

$$L(t) = L + \sum_{n=1}^{\infty} \frac{i^n}{n!} t^n [H, \ldots, [H, L] \ldots].$$

REMARK 1. Consider eq. (B.14). Let $|n\rangle$, $n = 0, 1, 2, \ldots$, be the eigenvectors of the Hamiltonian $H$ corresponding to the eigenvalue $E_n$, i.e. $H|n\rangle = E_n|n\rangle$. From (B.14) and (B.10) it follows that

$$\langle n|L(t)|n'\rangle = e^{i(E_n - E_{n'})} \langle n|L|n'\rangle. \tag{B.17}$$

Therefore, one can associate with each physical quantity a matrix of the form (B.17). The first formulation of quantum mechanics discovered by Heisenberg in 1925 was inspired by this observation. It is known nowadays as a *matrix quantum mechanics*. We point out that Heisenberg was not aware of the concept of an operator in a Hilbert space. ■

REMARK 2. It can be easily checked that whenever the observables $A$ and $B$ are related by

$$B = f(A),$$

where $f$ is analytic in $A$, then we have

$$B(t) = f(A(t)). \qquad \blacksquare$$

We now proceed to examine the time dependence of expectation values of observables. Let us assume for simplicity that the Hamiltonian is independent of time. Taking into account (B.15) and (B.14) we find that the expectation value of an observable $L$ satisfies the following equation:

$$i\frac{d\langle L\rangle_t}{dt} = \langle [L, H]\rangle_t, \qquad \langle L\rangle_{t=0} = \langle L\rangle. \tag{B.18}$$

From (B.18) we conclude that whenever $H$ and $L$ commute, i.e.,

$$[H, L] = 0, \tag{B.19}$$

then the expectation value at time $t = 0$ remains unchanged for all time. In analogy with classical mechanics we call $L$ satisfying (B.19) an *integral of the motion*.

Consider now a particle in an external field with a potential energy $V(\mathbf{q})$. The Newton equations describing the motion of such a particle are

$$\begin{aligned} \frac{d\mathbf{q}}{dt} &= \frac{\mathbf{p}}{m}, \\ \frac{d\mathbf{p}}{dt} &= -\frac{\partial V}{\partial \mathbf{q}} = \mathbf{F}(\mathbf{q}), \end{aligned} \tag{B.20}$$

where $\mathbf{q}$ and $\mathbf{p}$ is the position and momentum of the particle, $m$ is the mass of the particle and $\mathbf{F}$ is the force.

Evidently, the system (B.20) is Hamiltonian one, where the classical Hamiltonian is

$$H_c(\mathbf{q}, \mathbf{p}) = \frac{\mathbf{p}^2}{2m} + V(\mathbf{q}). \tag{B.21}$$

The quantal Hamiltonian corresponding to the classical system under investigation is given by

$$H = \frac{\hat{\mathbf{p}}^2}{2m} + V(\hat{\mathbf{q}}), \tag{B.22}$$

where $\hat{\mathbf{q}}$ and $\hat{\mathbf{p}}$ are the position and momentum operators, respectively (see appendix D).

Using (B.18), (D.1) and (D.3) we arrive at the following Ehrenfest equations:

$$\begin{aligned} \frac{d\langle\hat{\mathbf{q}}\rangle_t}{dt} &= \frac{\langle\hat{\mathbf{p}}\rangle_t}{m}, \\ \frac{d\langle\hat{\mathbf{p}}\rangle_t}{dt} &= -\left\langle\frac{\partial V}{\partial\hat{\mathbf{q}}}\right\rangle_t = \langle\mathbf{F}(\hat{\mathbf{q}})\rangle_t. \end{aligned}$$

We note that the Ehrenfest equations are very similar to the classical equations (B.20). Now let $|\psi(\mathbf{q})|^2$, where $\psi(\mathbf{q}) = \langle\mathbf{q}|\psi\rangle$ is the *wave function* of the particle and $|\mathbf{q}\rangle$ is the common eigenvector of the position operators, be the probability density for the coordinates (see (B.1b)). According to the *Ehrenfest's theorem*, if the probability density is approximately zero outside the convex region $\Omega$, where the force $\mathbf{F}$ is approximately constant, then $\langle\mathbf{F}(\hat{\mathbf{q}})\rangle \approx \mathbf{F}(\langle\hat{\mathbf{q}}\rangle)$ and the expectation values satisfy the classical equations of motion (B.20). It must be realized, however, that the assertions of the Ehrenfest's theorem cannot be valid at all time. Indeed, calculating the variance of the position operators we find that it is increasing in time. In physical literature, such a phenomenon is called the dispersion of wave-packets.

We end the discussion of quantum dynamics with a few remarks on the *interaction picture*. This picture is applied if the Hamiltonian is a sum of the Hamiltonian $H_0$ corresponding to free particles and the Hamiltonian $H_1(t)$ describing interaction of particles:

$$H(t) = H_0 + H_1(t).$$

Let us introduce the unitary operator of a "free evolution" such that

$$V(t) = e^{-itH_0}.$$

We postulate that in the interaction picture the time-dependent states are given by

$$\widetilde{|\phi, t\rangle} = V(t)^{-1}|\phi, t\rangle, \tag{B.23}$$

where $|\phi, t\rangle$ fulfils the Schrödinger equation (B.6).

Using (B.13) and proceeding analogously as with Heisenberg picture we find that the time-dependent observables in the interaction picture are defined as

$$\tilde{L}(t) = V(t)^{-1}LV(t). \tag{B.24}$$

By differentiating both sides of (B.23) we obtain the following equation satisfied by states in the interaction picture:

$$i\frac{d}{dt}\widetilde{|\phi, t\rangle} = \tilde{H}_1(t)\widetilde{|\phi, t\rangle}, \qquad \widetilde{|\phi, 0\rangle} = |\phi_0\rangle,$$

where

$$\tilde{H}_1(t) = V(t)^{-1}H_1(t)V(t)$$

is the Hamiltonian in the interaction picture.

On the other hand, taking the time derivative of (B.24) we arrive at the following equation which obey the time-dependent observables in the interaction picture:

$$i\frac{d\tilde{L}(t)}{dt} = [\tilde{L}(t), H_0], \qquad \tilde{L}(0) = L.$$

Note that the interaction picture coincides with the Heisenberg picture for $H(t) \equiv H_0$.

We conclude our brief exposition of fundamentals of quantum mechanics with a discussion of the basic properties of the density matrix. Let $p_k$ be the probability that the system is in the (normalized) state $|k\rangle$. Clearly, $\sum_k p_k = 1$. The Hermitian operator defined by

$$\rho = \sum_k p_k |k\rangle\langle k| \tag{B.25}$$

is called the *density matrix* or the statistical operator of the system. The matrix $\rho$ corresponds physically to a state in which we know that we are in the state $|k\rangle$ with probability $p_k$. Thus the system is in the state described by the density matrix if the information about the system is lesser than the maximal one provided by its state vector.

Now let $L$ be an observable. If the system is in the state $\rho$, then the probability that the result of a measurement of $L$ gives the value from a Borel set $\Delta$ is (see (B.1a)):

$$p_{L,\rho}(\Delta) = \sum_k p_k \langle k|E_L(\Delta)|k\rangle. \tag{B.26}$$

Hence, with the use of a resolution of the identity of the form (A.8) we get

$$p_{L,\rho}(\Delta) = \mathrm{Tr}(\rho E_L(\Delta)).$$

In particular, the probability of finding the system in the normalized state $|\phi\rangle$ is

$$p_\phi = \mathrm{Tr}(\rho|\phi\rangle\langle\phi|) = \langle\phi|\rho|\phi\rangle \geq 0. \tag{B.27}$$

Since (B.27) holds true for arbitrary state $|\phi\rangle$, we conclude that $\rho$ is nonnegative and we write

$$\rho \geq 0.$$

As an immediate consequence of (B.26) and (B.2), we find that the expectation value of the observable $L$ in the state $\rho$ is given by

$$\langle L\rangle_\rho = \mathrm{Tr}(\rho L). \tag{B.28}$$

On substituting in (B.28) $L = I$ we arrive at the following relation:

$$\mathrm{Tr}\rho = 1. \tag{B.29}$$

Suppose now that we know that the system is in a state $|\phi\rangle$. Referring back to (B.25), we find $p_k = \delta_{kl}$, $|l\rangle = |\phi\rangle$ and $\rho = |\phi\rangle\langle\phi|$. Obviously, we then have

$$\rho^2 = \rho.$$

Hence, using (B.29) we obtain

$$\mathrm{Tr}\rho^2 = 1. \tag{B.30}$$

When the condition (B.30) is valid, we say that the system is in a *pure state*. Clearly, the pure state is uniquely determined by a state vector (ray) in a Hilbert space. The states characterized by a statistical mixture (B.25), i.e. the density matrix, are referred to as the *mixed states*. Evidently,

$$\mathrm{Tr}\rho^2 < 1$$

for an arbitrary mixed state. The convenient indicator of the degree of purity of a state is given by the *entropy*, defined as [46]:

$$S = -\mathrm{Tr}(\rho \ln \rho).$$

In fact, it is easy to see from (B.30) that entropy is equal to zero for a pure state, while for a mixed state it is no longer zero.

According to the postulate that the time-dependent states satisfy the Schrödinger equation, the time-dependent density matrix is introduced by (see (B.7) and (B.10)):

$$\rho(t) = \sum_k p_k |k,t\rangle\langle k,t| = \sum_k p_k e^{-iHt}|k\rangle\langle k|e^{iHt} = e^{-iHt}\rho e^{iHt} \tag{B.31}$$

By differentiating the last formula of (B.31) we get

$$i\frac{d\rho(t)}{dt} = [H, \rho(t)], \qquad \rho(0) = \rho.$$

This equation is known as the von Neumann-Liouville equation.

REMARK. We point out that quantum mechanics specializes to the study of systems with a finite number of degrees of freedom. The quantum theory of systems with an infinite number of degrees of freedom is referred to as the *quantum field theory*. ■

APPENDIX C

# BOSE OPERATORS AND COHERENT STATES

## C.1 Bose operators and coherent states for $k$-degrees of freedom

In this appendix we outline the main facts about Bose operators and coherent states. We first discuss the case of the finite number of degrees of freedom. The *Bose creation* ($\mathbf{a}^\dagger$) and *annihilation* ($\mathbf{a}$) *operators*, where $\mathbf{a}^\dagger = (a_1^\dagger, \ldots, a_k^\dagger)$, $\mathbf{a} = (a_1, \ldots, a_k)$, satisfy the Heisenberg algebra

$$[a_i, a_j^\dagger] = \delta_{ij} I, \tag{C.1a}$$

$$[a_i, a_j] = [a_i^\dagger, a_j^\dagger] = 0, \qquad i, j = 1, \ldots, k, \tag{C.1b}$$

where $I$ is the identity operator.

Let us assume that the functions $\mathbf{f}, \mathbf{g}\colon \mathbf{R}^k \to \mathbf{R}^k$ are analytic so that the operator functions $\mathbf{f}(\mathbf{a})$, $\mathbf{g}(\mathbf{a}^\dagger)$ can be defined. Using (C.1) we find that

$$[f_i(\mathbf{a}), a_j^\dagger] = \frac{\partial f_i(\mathbf{a})}{\partial a_j}, \tag{C.2a}$$

$$[a_i, g_j(\mathbf{a}^\dagger)] = \frac{\partial g_j(\mathbf{a}^\dagger)}{\partial a_i^\dagger}, \qquad i, j = 1, \ldots, k. \tag{C.2b}$$

In general, we have

$$[F(\mathbf{a}), G(\mathbf{a}^\dagger)] = \sum_{\mathbf{n}\in \mathbf{Z}_+^k\setminus\{0\}} \left(\prod_{i=1}^{k} \frac{1}{n_i!}\right) \frac{\partial^{n_1}}{\partial a_1^{\dagger n_1}} \cdots \frac{\partial^{n_k}}{\partial a_k^{\dagger n_k}} G(\mathbf{a}^\dagger) \frac{\partial^{n_1}}{\partial a_1^{n_1}} \cdots \frac{\partial^{n_k}}{\partial a_k^{n_k}} F(\mathbf{a}), \tag{C.3}$$

where $F$ and $G$ are analytic in $\mathbf{a}$ and $\mathbf{a}^\dagger$, respectively; $\mathbf{Z}_+$ is the set of nonnegative integers.

The Hermitian operators $N_i = a_i^\dagger a_i$, $i = 1, \ldots, k$, are called the *number operators.* These operators satisfy

$$[N_i, N_j] = 0, \qquad [N_i, a_j] = -\delta_{ij} a_i, \qquad [N_i, a_j^\dagger] = \delta_{ij} a_i^\dagger, \qquad i, j = 1, \ldots, k. \tag{C.4}$$

The operator $N = \sum_{i=1}^{k} N_i$ is the *total number operator.* It follows that

$$[N, a_i] = -a_i, \qquad [N, a_i^\dagger] = a_i^\dagger, \qquad [N, \mathbf{a}^\dagger \cdot L\mathbf{a}] = 0, \qquad i = 1, \ldots, k, \tag{C.5}$$

where $\mathbf{u}\cdot\mathbf{v} = \sum_{i=1}^{k} u_i v_i$ and $L: \mathbf{R}^k \to \mathbf{R}^k$ is an arbitrary linear operator.

Now let $\mathcal{H}$ be a Hilbert space of states upon which the Bose operators act. We assume that there exists in $\mathcal{H}$ a unique normalized vector $|\mathbf{0}\rangle$ such that

$$\mathbf{a}|\mathbf{0}\rangle = \mathbf{0}, \tag{C.6a}$$

$$\langle\mathbf{0}|\mathbf{0}\rangle = 1. \tag{C.6b}$$

The vector $|\mathbf{0}\rangle$ is called the *vacuum vector*. We also assume that there is no nontrivial closed subspace of $\mathcal{H}$ which is invariant under the action of operators $\mathbf{a}$ and $\mathbf{a}^\dagger$. Let $|\mathbf{n}\rangle = |n_1, \ldots, n_k\rangle$, $\mathbf{n} \in \mathbf{Z}_+^k$, be the state vectors defined by

$$|\mathbf{n}\rangle = \left(\prod_{i=1}^{k} \frac{a_i^{\dagger n_i}}{\sqrt{n_i!}}\right) |\mathbf{0}\rangle. \tag{C.7}$$

These vectors are the common eigenvectors of the number operators, that is,

$$\mathbf{N}|\mathbf{n}\rangle = \mathbf{n}|\mathbf{n}\rangle. \tag{C.8}$$

It is easy to verify that the vectors (C.7) form an orthonormal basis of $\mathcal{H}$, i.e.,

$$\langle\mathbf{n}|\mathbf{n}'\rangle = \prod_{i=1}^{k} \delta_{n_i n_i'}, \tag{C.9}$$

$$\sum_{\mathbf{n}\in\mathbf{Z}_+^k} |\mathbf{n}\rangle\langle\mathbf{n}| = I. \tag{C.10}$$

In physical literature, the vectors $|\mathbf{n}\rangle$ are said to span the *occupation number representation.* The Hilbert space $\mathcal{H}$ constructed via (C.1), (C.6), (C.7) and (C.10) is known as the *Fock space with* k*-degrees of freedom.* The Bose operators act on the vectors $|\mathbf{n}\rangle$ as follows:

$$a_i|\mathbf{n}\rangle = \sqrt{n_i}\,|\mathbf{n} - \mathbf{e}_i\rangle, \tag{C.11a}$$

$$a_i^\dagger|\mathbf{n}\rangle = \sqrt{n_i + 1}\,|\mathbf{n} + \mathbf{e}_i\rangle, \qquad i = 1, \ldots, k, \tag{C.11b}$$

where $\mathbf{e}_i = (0, \ldots, 0, 1_i, 0, \ldots, 0)$ are the unit vectors.

We now collect some formulae corresponding to the case with $k = 1$. The Heisenberg algebra (C.1) can be then written as

$$[a, a^\dagger] = I. \tag{C.12}$$

Let $N = a^\dagger a$ be a number operator. The commutation relation (C.12) yields

$$a^r f(N) = f(N + r)a^r, \tag{C.13}$$

$$\frac{1}{r!} a^{\dagger r} a^r = \frac{1}{r!} \prod_{s=0}^{r-1} (N - s) = \binom{N}{r}, \tag{C.14}$$

$$\sum_{r=0}^{\infty} (\beta - 1)^r \binom{N}{r} = \beta^N, \tag{C.15}$$

where $f$ is analytic.

Evidently, the relations (C.11) reduce to

$$a|n\rangle = \sqrt{n}\,|n-1\rangle, \qquad a^\dagger|n\rangle = \sqrt{n+1}\,|n+1\rangle.$$

Hence we find

$$a^r|n\rangle = \left(\frac{n!}{(n-r)!}\right)^{\frac{1}{2}}|n-r\rangle, \qquad a^{\dagger r}|n\rangle = \left(\frac{(n+r)!}{n!}\right)^{\frac{1}{2}}|n+r\rangle.$$

REMARK. For $k = 3$, the state vectors $|n_1, n_2, n_3\rangle$ can be regarded as eigenvectors of the Hamiltonian for a three-dimensional harmonic oscillator. Indeed, on substituting in (B.21) $V(\mathbf{q}) = \frac{1}{2}m\sum_{i=1}^{3}\omega_i^2 q_i^2$ we arrive at the classical Hamiltonian for a three-dimensional oscillator such that

$$H_c = \frac{\mathbf{p}^2}{2m} + \frac{1}{2}m\sum_{i=1}^{3}\omega_i^2 q_i^2.$$

Hence, setting $m = \omega_i = 1$ we find that the quantal Hamiltonian (B.22) is

$$H = \frac{\hat{\mathbf{p}}^2}{2} + \frac{\hat{\mathbf{q}}^2}{2}. \tag{C.16}$$

Using (D.2a) and (C.1) we can express the Hamiltonian in terms of the Bose operators, to give

$$H = N + \frac{3}{2},$$

where $N = \sum_{i=1}^{3} N_i = \mathbf{a}^\dagger\cdot\mathbf{a}$ is the total number operator. Obviously,

$$H|n_1, n_2, n_3\rangle = (n + \tfrac{3}{2})|n_1, n_2, n_3\rangle,$$

where $n = \sum_{i=1}^{3} n_i$. Clearly, the state vectors $|n_1, \ldots, n_k\rangle$ can be treated as eigenvectors of the Hamiltonian for $k$ uncoupled one-dimensional harmonic oscillators. ∎

We now give the basic properties of the standard coherent states. The *coherent states* $|\mathbf{z}\rangle$, where $\mathbf{z} \in \mathbf{C}^k$, are usually introduced by

$$\mathbf{a}|\mathbf{z}\rangle = \mathbf{z}|\mathbf{z}\rangle, \tag{C.17}$$

that is they are eigenvectors of the Bose annihilation operators. The normalized coherent states can be defined as

$$|\mathbf{z}\rangle = D(\mathbf{z})|\mathbf{0}\rangle, \tag{C.18}$$

where $D(\mathbf{z})$ is a *displacement operator* given by

$$D(\mathbf{z}) = \exp(\mathbf{z}\cdot\mathbf{a}^\dagger - \mathbf{z}^*\cdot\mathbf{a}). \tag{C.19}$$

Here the asterisk denotes the complex conjugation.

Evidently, the displacement operator is unitary. Furthermore, using (B.16) and (C.1) we get

$$D^{-1}(\mathbf{z})\mathbf{a}D(\mathbf{z}) = \mathbf{a} + \mathbf{z}, \tag{C.20a}$$
$$D^{-1}(\mathbf{z})\mathbf{a}^{\dagger}D(\mathbf{z}) = \mathbf{a}^{\dagger} + \mathbf{z}^{*}. \tag{C.20b}$$

The equivalence of the definition (C.17) and (C.18) follows directly from (C.20a) and (C.6a). On applying the Baker-Hausdorff formula

$$e^{A+B} = e^{-\frac{1}{2}[A,B]}e^{A}e^{B},$$

for $[A,[A,B]] = [B,[A,B]] = 0$,

we obtain from (C.19) the following relations:

$$D(\mathbf{z}) = e^{-\frac{1}{2}|\mathbf{z}|^2}e^{\mathbf{z}\cdot\mathbf{a}^{\dagger}}e^{-\mathbf{z}^{*}\cdot\mathbf{a}}, \tag{C.21}$$
$$D(\mathbf{z})D(\mathbf{w}) = \exp[i\mathrm{Im}(\mathbf{z}\cdot\mathbf{w}^{*})]D(\mathbf{z}+\mathbf{w}), \tag{C.22}$$

where $|\mathbf{z}|^2 = \mathbf{z}\cdot\mathbf{z}^{*} = \sum_{i=1}^{k}|z_i|^2$.

In view of (C.22), it appears that the operators $D(\mathbf{z})$ form the unitary irreducible representation of the group $G$ with the multiplication law:

$$(t,\mathbf{z})(s,\mathbf{w}) = (t+s+\mathrm{Im}(\mathbf{z}\cdot\mathbf{w}^{*}), \mathbf{z}+\mathbf{w}).$$

The group $G$ is called the *Heisenberg-Weyl group.*

From (C.21) and (C.6a) we find that the normalized coherent states can be written in the following convenient form:

$$|\mathbf{z}\rangle = e^{-\frac{1}{2}|\mathbf{z}|^2}e^{\mathbf{z}\cdot\mathbf{a}^{\dagger}}|\mathbf{0}\rangle. \tag{C.23}$$

On taking into account (C.7), (C.17) and the formula

$$\langle\mathbf{0}|\mathbf{z}\rangle = e^{-\frac{1}{2}|\mathbf{z}|^2}, \tag{C.24}$$

which is an immediate consequence of (C.23), we arrive at the following relation describing the passage from the occupation number representation to the coherent-state representation:

$$\langle\mathbf{n}|\mathbf{z}\rangle = \left(\prod_{i=1}^{k}\frac{z_i^{n_i}}{\sqrt{n_i!}}\right)\exp(-\tfrac{1}{2}|\mathbf{z}|^2). \tag{C.25}$$

Hence, with the use of (C.10) we obtain the expansion of a normalized coherent state in the basis of the occupation number representation such that

$$|\mathbf{z}\rangle = e^{-\frac{1}{2}|\mathbf{z}|^2}\sum_{\mathbf{n}\in\mathbf{Z}_+^k}\left(\prod_{i=1}^{k}\frac{z_i^{n_i}}{\sqrt{n_i!}}\right)|\mathbf{n}\rangle.$$

The coherent states were discovered by Schrödinger in 1925 as the minimum uncertainty states. In fact, the Heisenberg relations (B.5), where $\hbar = 1$, become

$$\Delta\hat{q}_i \Delta\hat{p}_i = \frac{1}{2}, \qquad i = 1, 2, 3, \tag{C.26}$$

for arbitrary normalized coherent state. The relations (C.26) follows directly from (D.2a) and (B.3). In this sense, coherent states are closest to the classical ones. The denomination coherent states comes from quantum optics [47]. For a review of the numerous applications of coherent states in physics we refer the book [48].

REMARK. The standard coherent states (C.18) are the special case of the generalized coherent states [49] defined as follows. Let $U(g)$ be a unitary irreducible representation of a Lie group $G$ in a Hilbert space $\mathcal{H}$ and let $|\phi_0\rangle \in \mathcal{H}$ be a fixed vector. The set $\{|\phi_g\rangle\}$, $g \in G$, of generalized coherent states is introduced by

$$|\phi_g\rangle = U(g)|\phi_0\rangle. \tag{C.27}$$

The coherent states given by (C.27) are marked with points of the homogeneous space $G/H$, where $H$ is the stability subgroup of the state $|\phi_0\rangle$, that is $U(h)|\phi_0\rangle = e^{i\varphi}|\phi_0\rangle$ for every $h \in H$. ■

Using (C.18), (C.22) and (C.24) we find that the system of coherent states is not orthogonal one:

$$\langle \mathbf{z}|\mathbf{w}\rangle = \exp[-\tfrac{1}{2}(|\mathbf{z}|^2 + |\mathbf{w}|^2 - 2\mathbf{z}^*\cdot\mathbf{w})], \tag{C.28a}$$

$$|\langle \mathbf{z}|\mathbf{w}\rangle|^2 = \exp(-|\mathbf{z} - \mathbf{w}|^2). \tag{C.28b}$$

The nonorthogonal resolution of the identity for these vectors is given by

$$\int_{\mathbf{R}^{2k}} d\mu(\mathbf{z})\, |\mathbf{z}\rangle\langle\mathbf{z}| = I, \tag{C.29}$$

where

$$d\mu(\mathbf{z}) = \prod_{i=1}^{k} \frac{1}{\pi} d(\mathrm{Re}\, z_i)\, d(\mathrm{Im}\, z_i).$$

The coherent states form an overcomplete set. In fact, we have

$$|\mathbf{w}\rangle = \int_{\mathbf{R}^{2k}} d\mu(\mathbf{z})\, \langle\mathbf{z}|\mathbf{w}\rangle|\mathbf{z}\rangle \tag{C.30}$$

for arbitrary coherent state $|\mathbf{w}\rangle$. This means that coherent states are not independent and there exist complete subsystems of the family of coherent states $\{|\mathbf{z}\rangle\}$, where $\mathbf{z}$ runs over the whole $\mathbf{C}^k$.

We now return to eq. (C.25). It follows immediately from (C.25) that

$$|\langle \mathbf{n}|\mathbf{z}\rangle|^2 = \prod_{i=1}^{k} \frac{|z_i|^{2n_i}}{n_i!} e^{-|z_i|^2}.$$

This means that the probability of finding the system in the state $|\mathbf{n}\rangle$ when the system is actually in the coherent state $|\mathbf{z}\rangle$ has a Poisson distribution with a mean value $|z_i|^2$ of the random variable $n_i$. Therefore, the coherent states are also known as the *Poisson vectors* [50].

Suppose now that we are given an arbitrary state $|\phi\rangle \in \mathcal{H}$. Taking into account (C.10) and (C.25) we find that the function $\phi(\mathbf{z}^*) = \langle \mathbf{z}|\phi\rangle$ can be written as

$$\phi(\mathbf{z}^*) = \tilde{\phi}(\mathbf{z}^*)\exp(-\tfrac{1}{2}|\mathbf{z}|^2), \tag{C.31}$$

where $\tilde{\phi}(\mathbf{z}^*)$ is an analytic (entire) function.

On using (C.17), (C.24) and (C.31) we arrive at the following basis-independent form of (C.31):

$$|\phi\rangle = \tilde{\phi}(\mathbf{a}^\dagger)|\mathbf{0}\rangle. \tag{C.32}$$

Thus, it turns out that the vector $|\phi\rangle$ can be represented by an analytic (antianalytic) function $\tilde{\phi}(\mathbf{z}^*)$. In physical literature, such a representation is called the *Bargmann representation* [51]. The function $\tilde{\phi}(\mathbf{z}^*)$ is said to be the *symbol* of the vector $|\phi\rangle$. It follows directly from (C.29) and (C.31) that the inner product in the Bargmann representation is given by

$$\langle \phi|\psi\rangle = \int_{\mathbf{R}^{2k}} d\mu(\mathbf{z})\exp(-|\mathbf{z}|^2)(\tilde{\phi}(\mathbf{z}^*))^*\tilde{\psi}(\mathbf{z}^*). \tag{C.33}$$

The Hilbert space $F_2$ specified by the inner product (C.33) is known as the *Bargmann space* of entire functions [51]. One can easily check that the action of the Bose operators in the Bargmann representation has the following form:

$$\mathbf{a}\tilde{\phi}(\mathbf{z}^*) = \frac{\partial}{\partial \mathbf{z}^*}\tilde{\phi}(\mathbf{z}^*), \tag{C.34a}$$

$$\mathbf{a}^\dagger\tilde{\phi}(\mathbf{z}^*) = \mathbf{z}^*\tilde{\phi}(\mathbf{z}^*). \tag{C.34b}$$

Notice that the relations (C.34) can be written as

$$\langle \mathbf{z}|\mathbf{a}|\phi\rangle = \left(\frac{\partial}{\partial \mathbf{z}^*} + \frac{1}{2}\mathbf{z}\right)\langle \mathbf{z}|\phi\rangle,$$

$$\langle \mathbf{z}|\mathbf{a}^\dagger|\phi\rangle = \mathbf{z}^*\langle \mathbf{z}|\phi\rangle.$$

It should also be noted that the most effective way of calculating the inner product is to apply

$$\langle \phi|\psi\rangle = \tilde{\phi}^*(\tfrac{\partial}{\partial \mathbf{z}^*})\tilde{\psi}(\mathbf{z}^*)|_{\mathbf{z}^*=\mathbf{0}}, \tag{C.35}$$

where the function $\tilde{\phi}^*$ is defined by $(\tilde{\phi}^*(\mathbf{z}))^* = \tilde{\phi}(\mathbf{z}^*)$. The derivation of the relation (C.35) is straightforward, using eqs. (C.32), (C.3) and (C.6a).

Finally, we note that the Bargmann representation is spanned by the monomials (see (C.25)):

$$\tilde{\phi}_{\mathbf{n}}(\mathbf{z}^*) = \prod_{i=1}^{k} \frac{z_i^{*n_i}}{\sqrt{n_i!}}. \tag{C.36}$$

Obviously, the orthonormal basis (C.36) is the realization of the orthonormal complete set $\{|\mathbf{n}\rangle\}$ in the Bargmann representation. In particular, the vacuum vector $|\mathbf{0}\rangle$ is represented by

$$\tilde{\phi}_{\mathbf{0}}(\mathbf{z}^*) \equiv 1.$$

From (C.28a) it follows that the normalized coherent state $|\mathbf{w}\rangle$ is represented in the Bargmann representation by the function

$$\tilde{\phi}_{\mathbf{w}}(\mathbf{z}^*) = e^{-\frac{1}{2}|\mathbf{w}|^2} \exp(\mathbf{w}\cdot\mathbf{z}^*).$$

It is clear that this function is the solution of the following equation:

$$\frac{\partial}{\partial \mathbf{z}^*}\tilde{\phi}_{\mathbf{w}}(\mathbf{z}^*) = \mathbf{w}\tilde{\phi}_{\mathbf{w}}(\mathbf{z}^*)$$

which is the realization of the abstract equation

$$\mathbf{a}|\mathbf{w}\rangle = \mathbf{w}|\mathbf{w}\rangle$$

in the Bargmann representation.

REMARK. We note that the point spectrum of the Bose creation operators is empty. Indeed, writing the abstract equation

$$\mathbf{a}^\dagger|\boldsymbol{\lambda}\rangle = \boldsymbol{\lambda}|\boldsymbol{\lambda}\rangle$$

in the Bargmann representation, we get

$$(\mathbf{z}^* - \boldsymbol{\lambda})\tilde{\phi}_{\boldsymbol{\lambda}}(\mathbf{z}^*) = 0,$$

where $\tilde{\phi}_{\boldsymbol{\lambda}}(\mathbf{z}^*) = \langle \mathbf{z}|\boldsymbol{\lambda}\rangle \exp(\frac{1}{2}|\mathbf{z}|^2)$. Using the fact that $\tilde{\phi}_{\boldsymbol{\lambda}}(\mathbf{z}^*)$ is entire function of $\mathbf{z}^*$ we conclude that $\tilde{\phi}_{\boldsymbol{\lambda}}(\mathbf{z}^*) \equiv 0$. Therefore, $|\boldsymbol{\lambda}\rangle = 0$. ■

We now return to eq. (C.30). Projecting both sides of (C.30) onto the state $|\phi\rangle$ and using (C.28a) and (C.31) we arrive at the following relation:

$$\tilde{\phi}(\mathbf{w}^*) = \int_{\mathbf{R}^{2k}} d\mu(\mathbf{z}) \exp(-|\mathbf{z}|^2) \exp(\mathbf{w}^*\cdot\mathbf{z})\tilde{\phi}(\mathbf{z}^*). \tag{C.37}$$

Recall that if there exists in a Hilbert space $\mathcal{H}$ of complex-valued functions on the set $X$, the function $\mathcal{K}$: $X \times X \to \mathbf{C}$, satisfying

$$\mathcal{K}(\cdot, x) \in \mathcal{H} \quad \text{for all} \quad x \in X,$$
$$\phi(x) = \langle \mathcal{K}(\cdot, x)|\phi\rangle \quad \text{for all} \quad \phi \in \mathcal{H},$$

then $\mathcal{K}$ is called the *Bergman reproducing kernel* [52]. Thus the reproducing kernel in the Bargmann space is given by

$$\mathcal{K}(\mathbf{w}^*, \mathbf{z}) = \exp(\mathbf{w}^* \cdot \mathbf{z}).$$

On taking the Hermitian conjugate of (C.37) we find that the reproducing property can be written in the following form:

$$\tilde{\psi}(\mathbf{w}) = \int_{\mathbf{R}^{2k}} d\mu(\mathbf{z}) \exp(-|\mathbf{z}|^2) \exp(\mathbf{w} \cdot \mathbf{z}^*) \tilde{\psi}(\mathbf{z}). \tag{C.38}$$

We now study the operators in the coherent-state representation. An important property of coherent states is that they allow to associate to each operator a function which entirely determines this operator. Such a function is called a *symbol* of the operator. Let $L$ be a linear operator. The *covariant symbol* $L(\mathbf{z}^*, \mathbf{z})$ of the operator $L$ is defined by [53]:

$$L(\mathbf{z}^*, \mathbf{z}) = \langle \mathbf{z} | L | \mathbf{z} \rangle, \tag{C.39}$$

where $|\mathbf{z}\rangle$ is a normalized coherent state.

An immediate consequence of (C.39) is the following simple formula on the trace of the operator $L$:

$$\mathrm{Tr} L = \int_{\mathbf{R}^{2k}} d\mu(\mathbf{z})\, L(\mathbf{z}^*, \mathbf{z}). \tag{C.40}$$

Suppose now that the operator $L$ can be written in the form

$$L = \sum_{\mathbf{n},\mathbf{m} \in \mathbf{Z}_+^k} c_{\mathbf{nm}} \left( \prod_{i=1}^{k} a_i^{\dagger n_i} \right) \left( \prod_{j=1}^{k} a_j^{m_j} \right) \tag{C.41}$$

which is referred to as the *normal* or *Wick form*. Evidently, the symbol of the operator (C.41) is given by

$$L(\mathbf{z}^*, \mathbf{z}) = \sum_{\mathbf{n},\mathbf{m} \in \mathbf{Z}_+^k} c_{\mathbf{nm}} \prod_{i=1}^{k} z_i^{*n_i} z_i^{m_i}.$$

It thus appears that the symbol (C.39) of the operator (C.41) enables one to determine the coefficients $c_{\mathbf{nm}}$ in the normal form of the operator $L$.

Finally, we examine the action of operators in the Bargmann representation. On taking into account (C.29), (C.10) and (C.31) we arrive at the following relation:

$$(L\tilde{\phi})(\mathbf{w}^*) = \int_{\mathbf{R}^{2k}} d\mu(\mathbf{z})\, e^{-|\mathbf{z}|^2} \mathcal{L}(\mathbf{w}^*, \mathbf{z}) \tilde{\phi}(\mathbf{z}^*). \tag{C.42}$$

This means that an arbitrary operator $L$ is represented in the Bargmann representation by an integral operator. The kernel $\mathcal{L}(\mathbf{w}^*, \mathbf{z})$ of the integral operator (C.42) is

$$\mathcal{L}(\mathbf{w}^*, \mathbf{z}) = \sum_{\mathbf{n},\mathbf{m} \in \mathbf{Z}_+^k} L_{\mathbf{nm}} \prod_{i=1}^{k} \frac{w_i^{*n_i} z_i^{m_i}}{\sqrt{n_i! m_i!}}, \tag{C.43}$$

where $L_{\mathbf{nm}} = \langle \mathbf{n}|L|\mathbf{m}\rangle$.

Clearly, the kernel (C.43) can be written as

$$\mathcal{L}(\mathbf{w}^*, \mathbf{z}) = \exp[\tfrac{1}{2}(|\mathbf{w}|^2 + |\mathbf{z}|^2)]\langle \mathbf{w}|L|\mathbf{z}\rangle. \tag{C.44}$$

Hence we find that the kernel $\mathcal{L}$ is related to the covariant symbol of the operator $L$ by

$$\mathcal{L}(\mathbf{z}^*, \mathbf{z}) = e^{|\mathbf{z}|^2} L(\mathbf{z}^*, \mathbf{z}).$$

On the other hand, using (C.44) and (C.29) we obtain

$$\mathcal{C}(\mathbf{w}^*, \mathbf{z}) = \int_{\mathbf{R}^{2k}} d\mu(\mathbf{z}')\, e^{-|\mathbf{z}'|^2} \mathcal{A}(\mathbf{w}^*, \mathbf{z}')\mathcal{B}(\mathbf{z}'^*, \mathbf{z}),$$

where $C = AB$, that is the product of operators $A$ and $B$ is represented in the Bargmann representation by the convolution of their corresponding kernels $\mathcal{A}$ and $\mathcal{B}$.

## C.2 Bose field operators and functional coherent states

We now summarize the basic facts about Bose field operators and functional coherent states. The *Bose field creation* ($\mathbf{a}^\dagger(x)$) and *annihilation* ($\mathbf{a}(x)$) *operators*, where $\mathbf{a}^\dagger(x) = (a_1^\dagger(x), \ldots, a_k^\dagger(x))$ and $\mathbf{a}(x) = (a_1(x), \ldots, a_k(x))$, $x \in \mathbf{R}^s$, satisfy the following canonical commutation relations:

$$[a_i(x), a_j^\dagger(x')] = \delta_{ij}\delta(x - x')I, \tag{C.45a}$$

$$[a_i(x), a_j(x')] = [a_i^\dagger(x), a_j^\dagger(x')] = 0, \qquad x,\, x' \in \mathbf{R}^s, \qquad i,\, j = 1, \ldots, k, \tag{C.45b}$$

where $\delta(x - x') = \prod_{r=1}^{s} \delta(x_r - x_r')$. We note that the Bose field operators are actually operator-valued distributions. Let $\mathbf{f}(\mathbf{a}(x)) \equiv \mathbf{f}(\mathbf{a}(x), D^\alpha \mathbf{a}(x))$ and $\mathbf{g}(\mathbf{a}(x)) \equiv \mathbf{g}(\mathbf{a}(x), D^\beta \mathbf{a}(x))$ with $D^\gamma \mathbf{a} = (D^{\gamma_1} a_1, \ldots, D^{\gamma_k} a_k)$; $D^\rho = \partial^{|\rho|}/\partial x_1^{\rho_1} \cdots \partial x_s^{\rho_s}$, be analytic operator functions. The commutation relations (C.45) give

$$[f_i(\mathbf{a}(x)), a_j^\dagger(x')] = \frac{\partial f_i(\mathbf{a}(x))}{\partial a_j(x)}\delta(x - x') + \sum_{\alpha_j} \frac{\partial f_i(\mathbf{a}(x))}{\partial D^{\alpha_j} a_j(x)} D^{\alpha_j}\delta(x - x'), \tag{C.46a}$$

$$[a_i(x), g_j(\mathbf{a}^\dagger(x'))] = \frac{\partial g_j(\mathbf{a}^\dagger(x'))}{\partial a_i^\dagger(x')}\delta(x - x') + \sum_{\beta_i} \frac{\partial g_j(\mathbf{a}^\dagger(x'))}{\partial D^{\beta_i} a_i^\dagger(x')} D^{\beta_i}\delta(x - x'), \tag{C.46b}$$

where $i, j = 1, \ldots, k$, and $D^{\alpha_j}$, $D^{\beta_i}$ act on the $x$ and $x'$ variable in $\delta(x - x')$, respectively.

Suppose that we are given an operator functional $I[\mathbf{a}]$:

$$I[\mathbf{a}] = \int d^s x\, \mathcal{I}(\mathbf{a}),$$

where $\mathcal{I}(\mathbf{a}) \equiv \mathcal{I}(\mathbf{a}, D^\sigma \mathbf{a})$ is analytic in $\mathbf{a}$, $D^\sigma \mathbf{a}$.

Eqs. (E.4) and (C.45a) taken together yield

$$[I[\mathbf{a}], \mathbf{a}^\dagger(x)] = \frac{\delta I[\mathbf{a}]}{\delta \mathbf{a}(x)}, \tag{C.47}$$

where $\delta/\delta\mathbf{a}(x)$ designates the functional derivative (see appendix E).

The Hermitian number operators such that

$$N_i = \int d^s x \, a_i^\dagger(x) a_i(x), \qquad i = 1, \ldots, k,$$

obey the following relations:

$$[N_i, N_j] = 0, \qquad [N_i, a_j(x)] = -\delta_{ij} a_i(x), \qquad [N_i, a_j^\dagger(x)] = \delta_{ij} a_i^\dagger(x),$$

where $i, j = 1, \ldots, k$.

As in the finite number of degrees of freedom set-up, we assume that there exists in the Hilbert space $\mathcal{H}$ of states, where the Bose field operators act, a unique normalized vacuum vector $|\mathbf{0}\rangle$ satisfying

$$\mathbf{a}(x)|\mathbf{0}\rangle = \mathbf{0} \quad \text{for all} \quad x \in \mathbf{R}^s. \tag{C.48}$$

We remark that (C.48) should be regarded as an equality between distributions. More precisely, eq. (C.48) means that

$$\int d^s x \, \boldsymbol{\phi}(x) \cdot \mathbf{a}(x) |\mathbf{0}\rangle = 0,$$

where $\boldsymbol{\phi}$ is an element of a suitable test-function space.

Furthermore, we assume that there is no nontrivial closed subspace of $\mathcal{H}$ which is invariant under the action of operators $\mathbf{a}(x)$ and $\mathbf{a}^\dagger(x)$. We now define the state vectors by

$$|x_{11}, \ldots, x_{1n_1}, \ldots, x_{k1}, \ldots, x_{kn_k}\rangle = \left( \prod_{i=1}^{k} \prod_{j=1}^{n_i} a_i^\dagger(x_{ij}) \right) |\mathbf{0}\rangle. \tag{C.49}$$

We note that the vectors (C.49) are actually vector-valued distributions. They are the common eigenvectors of the number operators, i.e.,

$$\mathbf{N}|x_{11}, \ldots, x_{1n_1}, \ldots, x_{k1}, \ldots, x_{kn_k}\rangle = \mathbf{n}|x_{11}, \ldots, x_{1n_1}, \ldots, x_{k1}, \ldots, x_{kn_k}\rangle.$$

These vectors form an orthogonal and complete set, namely,

$$\begin{aligned} &\langle x_{11}, \ldots, x_{1n_1}, \ldots, x_{k1}, \ldots, x_{kn_k} | x'_{11}, \ldots, x'_{1n'_1}, \ldots, x'_{k1}, \ldots, x'_{kn'_k}\rangle \\ &= \left( \prod_{i=1}^{k} \delta_{n_i n'_i} \right) \sum_{\sigma_1, \ldots, \sigma_k} \prod_{i=1}^{k} \prod_{j=1}^{n_i} \delta(x_{ij} - x'_{i\sigma_i(j)}), \end{aligned} \tag{C.50}$$

where $\sigma_i$'s are the permutations of the set $\{1,\ldots,n_i\}$, and

$$\sum_{\mathbf{n}\in\mathbf{Z}_+^k}\left(\prod_{i=1}^{k}\frac{1}{n_i!}\right)\int\prod_{r=1}^{k}\prod_{j=1}^{n_r}d^s x_{rj}\,|x_{11},\ldots,x_{1n_1},\ldots,x_{k1},\ldots,x_{kn_k}\rangle$$
$$\times\langle x_{11},\ldots,x_{1n_1},\ldots,x_{k1},\ldots,x_{kn_k}|=I. \tag{C.51}$$

The vectors (C.49) are said to span the *coordinate representation*. The Hilbert space determined via (C.45), (C.48), (C.49) and (C.51) is referred to as the boson Fock space (see remark below). The Bose operators act on the vectors (C.49) in the following way:

$$a_i(x)|x_{11},\ldots,x_{1n_1},\ldots,x_{k1},\ldots,x_{kn_k}\rangle$$
$$=\sum_{j=1}^{n_i}\delta(x-x_{ij})|x_{11},\ldots,x_{1n_1},\ldots,\check{x}_{ij},\ldots,x_{k1},\ldots,x_{kn_k}\rangle, \tag{C.52a}$$
$$a_i^\dagger(x)|x_{11},\ldots,x_{1n_1},\ldots,x_{k1},\ldots,x_{kn_k}\rangle$$
$$=|x_{11},\ldots,x_{1n_1},\ldots,x_{i1},\ldots,x_{in_i},x,\ldots,x_{k1},\ldots,x_{kn_k}\rangle, \tag{C.52b}$$

where the reversed hat over $x_{ij}$ denotes that this variable should be omitted from the set $\{x_{i1},\ldots,x_{in_i}\}$.

Having in mind the applications of the boson calculus in the theory of nonlinear partial differential equations introduced in chapter 4 we now rewrite some of the above formulae for $k=1$. The algebra (C.45) becomes

$$[a(x),a^\dagger(x')]=\delta(x-x')I, \tag{C.53a}$$
$$[a(x),a(x')]=[a^\dagger(x),a^\dagger(x')]=0,\qquad x,\,x'\in\mathbf{R}^s. \tag{C.53b}$$

The basis vectors of the coordinate representation defined by

$$|x_1,\ldots,x_n\rangle=\left(\prod_{i=1}^{n}a^\dagger(x_i)\right)|0\rangle, \tag{C.54}$$

are the eigenvectors of the number operator $N=\int d^s x\,a^\dagger(x)a(x)$, i.e.,

$$N|x_1,\ldots,x_n\rangle=n|x_1,\ldots,x_n\rangle.$$

They satisfy the following orthogonality relation:

$$\langle x_1,\ldots,x_n|x'_1,\ldots,x'_{n'}\rangle=\delta_{nn'}\sum_{\sigma}\prod_{i=1}^{n}\delta(x_i-x'_{\sigma(i)}), \tag{C.55}$$

where $\sigma$ is a permutation of the set $\{1,\ldots,n\}$, and the completeness relation

$$\sum_{n}\frac{1}{n!}\int d^s x_1\cdots d^s x_n\,|x_1,\ldots,x_n\rangle\langle x_1,\ldots,x_n|=I. \tag{C.56}$$

The action of the Bose operators $a(x)$, $a^\dagger(x)$ on the vectors (C.54) has the following form:

$$a(x)|x_1,\ldots,x_n\rangle = \sum_{i=1}^{n} \delta(x - x_i)|x_1,\ldots,\check{x}_i,\ldots,x_n\rangle, \tag{C.57a}$$

$$a^\dagger(x)|x_1,\ldots,x_n\rangle = |x_1,\ldots,x_n,x\rangle, \tag{C.57b}$$

where the reversed hat over $x_i$ denotes that this variable should be omitted from the set $\{x_1,\ldots,x_n\}$.

REMARK. For $s = 3$, the Bose creation (annihilation) operator $a_i^\dagger(x)$ $(a_i(x))$ is said to create (annihilate) a particle (boson) of the kind $i$ at the point $x \in \mathbf{R}^3$. The vector $|x_{11},\ldots,x_{1n_1},\ldots,x_{k1},\ldots,x_{kn_k}\rangle$ represents the state in which there are $n_i$ identical particles of the kind $i$ with coordinates $x_{ij} \in \mathbf{R}^3$, $j = 1,\ldots,n_i$, respectively, where $i = 1,\ldots,k$. The vacuum vector $|0\rangle$ represents the state in which there are no particles.

We now consider, for simplicity, the case with $k = 1$. In view of (C.56), an arbitrary vector $|\phi\rangle \in \mathcal{H}$ can be written as

$$|\phi\rangle = \sum_n \frac{1}{n!}\int d^s x_1 \cdots d^s x_n\, \phi_n(x_1,\ldots,x_n)|x_1,\ldots,x_n\rangle,$$

where $\phi_n(x_1,\ldots,x_n) = \langle x_1,\ldots,x_n|\phi\rangle$.

This means that the Hilbert space $\mathcal{H}$ is the direct sum of eigenspaces $\mathcal{H}_n$ spanned by eigenvectors of the number operator $N$ corresponding to the eigenvalue $n$, that is $\mathcal{H} = \bigoplus_{n=0}^{\infty} \mathcal{H}_n$ and

$$|\psi\rangle \in \mathcal{H}_n \quad \text{if and only if} \quad N|\psi\rangle = n|\psi\rangle.$$

By virtue of (C.54) and (C.53b) the functions $\phi_n(x_1,\ldots,x_n) = \langle x_1,\ldots,x_n|\phi\rangle$ are invariant with respect to an arbitrary permutation of their arguments. Therefore, the matrix element $\langle x_1,\ldots,x_n|\phi\rangle$ can be identified with the product

$$\langle x_n|\cdots\langle x_1|\sum_\sigma U(\sigma)|\phi_1\rangle\cdots|\phi_n\rangle,$$

where $\sigma$ is a permutation of the set $\{1,\ldots,n\}$ and the linear, unitary operator $U(\sigma)$ is defined by

$$U(\sigma)|\phi_1\rangle\cdots|\phi_n\rangle = |\phi_{\sigma(1)}\rangle\cdots|\phi_{\sigma(n)}\rangle.$$

Using the fact that the inner product in $\mathcal{H}$ is

$$\langle\phi|\psi\rangle = \sum_n \frac{1}{n!}\int d^s x_1 \cdots d^s x_n\, \phi_n^*(x_1,\ldots,x_n)\psi_n(x_1,\ldots,x_n),$$

we can write

$$\mathcal{H} = \bigoplus_{n=0}^{\infty} S_n L^2(\mathbf{R}^s, d^s x)^{\otimes n},$$

where $L^2(\mathbf{R}^s, d^s x)^{\otimes n}$ denotes the full tensor product of $n$ copies of $L^2(\mathbf{R}^s, d^s x)$ (complex Hilbert space of square integrable functions) for $n \geq 1$, and is defined as $\mathbf{C}$ if $n = 0$; the linear operator $S_n$ is given by

$$S_n = \frac{1}{n!} \sum_\sigma U(\sigma).$$

The Hilbert space $\mathcal{H}$ is called the boson Fock space over $L^2(\mathbf{R}^s, d^s x)$. In general, the Hilbert space

$$\mathcal{F}_s(\mathcal{H}) = \bigoplus_{n=0}^{\infty} S_n \mathcal{H}^{\otimes n}.$$

is referred to as the *boson Fock space over* $\mathcal{H}$. The space $S_n \mathcal{H}^{\otimes n}$ is called the $n$-fold symmetric tensor product of $\mathcal{H}$. In physical literature, the space $S_n \mathcal{H}^{\otimes n}$ is known as the n-*particle subspace* of $\mathcal{F}_s(\mathcal{H})$. ∎

We end this appendix by outlining some of the most important properties of the functional coherent states. As with a finite number of degrees of freedom set-up, the *functional coherent states* $|\boldsymbol{\xi}\rangle$, where $\boldsymbol{\xi} \in \bigoplus_{i=1}^k L^2(\mathbf{R}^s, d^s x)$, are defined as the eigenvectors of the Bose annihilation operators, that is,

$$\mathbf{a}(x)|\boldsymbol{\xi}\rangle = \boldsymbol{\xi}(x)|\boldsymbol{\xi}\rangle. \tag{C.58}$$

The normalized functional coherent states can be introduced by

$$|\boldsymbol{\xi}\rangle = \exp\left(-\frac{1}{2}\int d^s x\, |\boldsymbol{\xi}|^2\right) \exp\left(\int d^s x\, \boldsymbol{\xi}(x)\cdot \mathbf{a}^\dagger(x)\right) |\mathbf{0}\rangle. \tag{C.59}$$

The set of functional coherent states is not orthogonal, namely,

$$\langle \boldsymbol{\xi}|\boldsymbol{\eta}\rangle = \exp\left[-\frac{1}{2}\int d^s x\, (|\boldsymbol{\xi}|^2 + |\boldsymbol{\eta}|^2 - 2\boldsymbol{\xi}^*\cdot\boldsymbol{\eta})\right]. \tag{C.60}$$

The formal resolution of the identity for these states can be expressed as

$$\int_{\Omega^{2k}} \mathrm{D}^2\boldsymbol{\xi}\, |\boldsymbol{\xi}\rangle\langle\boldsymbol{\xi}| = I, \tag{C.61}$$

where $\Omega$ is the real space $\mathcal{S}'(\mathbf{R}^s)$ of the tempered distributions or the real space $\mathcal{D}'(\mathbf{R}^s)$ of Schwartz distributions,

$$\mathrm{D}^2\boldsymbol{\xi} = \prod_{i=1}^{k} \mathrm{D}(\mathrm{Re}\,\xi_i)\mathrm{D}(\mathrm{Im}\,\xi_i),$$

and the formal expression

$$\langle 0|u\rangle^2 \mathrm{D}u = \exp\left(-\int d^s x\, u^2\right) \mathrm{D}u,$$

where $u \in L^2_{\mathbf{R}}(\mathbf{R}^s, d^s x)$ (real Hilbert space of square integrable functions), designates the Gaussian measure [54] $d\mu(u)$ on $\Omega$.

The passage from the coordinate representation to the functional coherent-state representation is described by

$$\langle x_{11}, \ldots, x_{1n_1}, \ldots, x_{k1}, \ldots, x_{kn_k}|\boldsymbol{\xi}\rangle = \left(\prod_{i=1}^{k}\prod_{j=1}^{n_i} \xi_i(x_{ij})\right) \exp\left(-\frac{1}{2}\int d^s x\, |\boldsymbol{\xi}|^2\right). \quad \text{(C.62)}$$

Clearly, for $k = 1$, this relation reduces to (see (C.54)):

$$\langle x_1, \ldots, x_n|\xi\rangle = \left(\prod_{i=1}^{n} \xi(x_i)\right) \exp\left(-\frac{1}{2}\int d^s x\, |\xi|^2\right). \quad \text{(C.63)}$$

Now let $|\phi\rangle \in \mathcal{H}$ be an arbitrary state. Using (C.51) and (C.62) we see that the functional $\phi[\boldsymbol{\xi}^*] = \langle\boldsymbol{\xi}|\phi\rangle$ can be written as

$$\phi[\boldsymbol{\xi}^*] = \tilde{\phi}[\boldsymbol{\xi}^*] \exp\left(-\frac{1}{2}\int d^s x\, |\boldsymbol{\xi}|^2\right), \quad \text{(C.64)}$$

where the functional $\tilde{\phi}[\boldsymbol{\xi}^*]$ is analytic.

On taking into account (C.58) and (C.60) we obtain the following abstract form of (C.64):

$$|\phi\rangle = \tilde{\phi}[\mathbf{a}^\dagger]|\mathbf{0}\rangle. \quad \text{(C.65)}$$

From (C.61) and (C.64) it follows that

$$\langle\phi|\psi\rangle = \int_{\Omega^{2k}} \mathrm{D}^2\boldsymbol{\xi} \exp\left(-\int d^s x\, |\boldsymbol{\xi}|^2\right) (\tilde{\phi}[\boldsymbol{\xi}^*])^* \tilde{\psi}[\boldsymbol{\xi}^*]. \quad \text{(C.66)}$$

Such a realization of the Hilbert space $\mathcal{H}$ is called the *functional Bargmann representation* [55]. The action of the Bose operators in this representation is

$$\mathbf{a}(x)\tilde{\phi}[\boldsymbol{\xi}^*] = \frac{\delta}{\delta\boldsymbol{\xi}^*(x)}\tilde{\phi}[\boldsymbol{\xi}^*], \quad \text{(C.67a)}$$

$$\mathbf{a}^\dagger(x)\tilde{\phi}[\boldsymbol{\xi}^*] = \boldsymbol{\xi}^*(x)\tilde{\phi}[\boldsymbol{\xi}^*], \quad \text{(C.67b)}$$

where $\delta/\delta\boldsymbol{\xi}^*(x)$ designates the functional derivative (see appendix E).

Suppose that we are given an operator $L$ of trace class. We have

$$\mathrm{Tr}L = \int_{\Omega^{2k}} \mathrm{D}^2\boldsymbol{\xi}\, L[\boldsymbol{\xi}^*, \boldsymbol{\xi}] \quad \text{(C.68a)}$$

$$= \int_{\Omega^{2k}} \mathrm{D}^2\boldsymbol{\xi} \exp\left(-\int d^s x\, |\boldsymbol{\xi}|^2\right) \mathcal{L}[\boldsymbol{\xi}^*, \boldsymbol{\xi}], \quad \text{(C.68b)}$$

where the functional $L[\boldsymbol{\xi}^*,\boldsymbol{\xi}] = \langle\boldsymbol{\xi}|L|\boldsymbol{\xi}\rangle$ is the covariant symbol of the operator $L$ and we assume that $L$ is written in the normal form. By this we mean the expansion analogous to (C.41), where the Bose creation operators are placed on the left-hand side of the Bose annihilation operators. The functional $\mathcal{L}[\boldsymbol{\xi}^*,\boldsymbol{\xi}]$ given by

$$\mathcal{L}[\boldsymbol{\xi}^*,\boldsymbol{\xi}] = \exp\left(\int d^s x\, |\boldsymbol{\xi}|^2\right) L[\boldsymbol{\xi}^*,\boldsymbol{\xi}], \tag{C.69}$$

is said to correspond to the matrix form $\mathcal{L}_{\mathbf{nn'}}$ of the operator $L$ such that

$$\mathcal{L}_{\mathbf{nn'}} = \langle x_{11},\ldots,x_{1n_1},\ldots,x_{k1},\ldots,x_{kn_k}|L|x'_{11},\ldots,x'_{1n'_1},\ldots,x'_{k1},\ldots,x'_{kn'_k}\rangle. \tag{C.70}$$

Indeed, using (C.51), (C.62) and (C.70) we easily obtain (C.69).

Now let $L$ be a bounded operator. The action of the operator $L$ in the functional Bargmann representation is described by

$$(L\tilde{\phi})[\boldsymbol{\eta}^*] = \int_{\Omega^{2k}} \mathrm{D}^2\boldsymbol{\xi} \exp\left(-\int d^s x\, |\boldsymbol{\xi}|^2\right) \mathcal{L}[\boldsymbol{\eta}^*,\boldsymbol{\xi}]\tilde{\phi}[\boldsymbol{\xi}^*]. \tag{C.71}$$

The functional $\mathcal{C}$ corresponding to the product $C = AB$ of bounded operators $A$ and $B$ is

$$\mathcal{C}[\boldsymbol{\eta}^*,\boldsymbol{\xi}] = \int_{\Omega^{2k}} \mathrm{D}^2\boldsymbol{\xi}' \exp\left(-\int d^s x\, |\boldsymbol{\xi}'|^2\right) \mathcal{A}[\boldsymbol{\eta}^*,\boldsymbol{\xi}']\mathcal{B}[\boldsymbol{\xi}'^*,\boldsymbol{\xi}].$$

REMARK. The mathematical formalism of the field operators is known as the *method of second quantization*. The reader who is interested in a more detailed analysis of the field operators and functional coherent states is referred to Berezin treatise [56]. ■

APPENDIX D

# POSITION AND MOMENTUM OPERATORS

Our aim in this appendix is to introduce the basic properties of the position and momentum operators. The *position* ($\hat{\mathbf{q}}$) and *momentum* ($\hat{\mathbf{p}}$) *operators*, where $\hat{\mathbf{q}} = (\hat{q}_1, \ldots, \hat{q}_k)$ and $\hat{\mathbf{p}} = (\hat{p}_1, \ldots, \hat{p}_k)$, obey the canonical commutation relations

$$[\hat{q}_r, \hat{p}_s] = i\delta_{rs}I, \tag{D.1a}$$

$$[\hat{q}_r, \hat{q}_s] = [\hat{p}_r, \hat{p}_s] = 0, \qquad r,\, s = 1, \ldots, k, \tag{D.1b}$$

where we set $\hbar I = I$.

The position and momentum operators are linked to the Bose operators by

$$\hat{\mathbf{q}} = \frac{1}{\sqrt{2}}(\mathbf{a} + \mathbf{a}^\dagger), \qquad \hat{\mathbf{p}} = \frac{i}{\sqrt{2}}(\mathbf{a}^\dagger - \mathbf{a}), \tag{D.2a}$$

$$\mathbf{a} = \frac{1}{\sqrt{2}}(\hat{\mathbf{q}} + i\hat{\mathbf{p}}), \qquad \mathbf{a}^\dagger = \frac{1}{\sqrt{2}}(\hat{\mathbf{q}} - i\hat{\mathbf{p}}). \tag{D.2b}$$

It is clear that position and momentum operators are Hermitian. Let $\mathbf{f}, \mathbf{g}\colon \mathbf{R}^k \to \mathbf{R}^k$ be analytic functions. We introduce the operator functions $\mathbf{f}(\hat{\mathbf{q}})$ and $\mathbf{g}(\hat{\mathbf{p}})$. From (D.1) it follows that

$$[f_r(\hat{\mathbf{q}}), \hat{p}_s] = i\frac{\partial f_r(\hat{\mathbf{q}})}{\partial \hat{q}_s}, \tag{D.3a}$$

$$[\hat{q}_r, g_s(\hat{\mathbf{p}})] = i\frac{\partial g_s(\hat{\mathbf{p}})}{\partial \hat{p}_r}, \qquad r,\, s = 1, \ldots, k. \tag{D.3b}$$

Consider now the eigenvectors $|\mathbf{q}\rangle$, where $\mathbf{q} \in \mathbf{R}^k$, of the position operators:

$$\hat{\mathbf{q}}|\mathbf{q}\rangle = \mathbf{q}|\mathbf{q}\rangle.$$

For $k = 3$, the vector $|\mathbf{q}\rangle$ represents the state in which a particle is localized at the point $\mathbf{q} \in \mathbf{R}^3$. The normalized eigenvectors are given by

$$|\mathbf{q}\rangle = \pi^{-\frac{k}{4}} \exp(\tfrac{1}{2}\mathbf{q}^2)\exp[-\tfrac{1}{2}(\mathbf{a}^\dagger - \sqrt{2}\,\mathbf{q})^2]|\mathbf{0}\rangle.$$

The eigenvectors of the position operators are orthogonal. We have

$$\langle \mathbf{q}|\mathbf{q}'\rangle = \delta(\mathbf{q} - \mathbf{q}').$$

These vectors form the complete set. The resolution of the identity is of the form

$$\int d^k q\, |\mathbf{q}\rangle\langle\mathbf{q}| = I. \tag{D.4}$$

The completeness gives rise to a functional representation of vectors such that

$$\langle\phi|\psi\rangle = \int d^k q\, \phi^*(\mathbf{q})\psi(\mathbf{q}),$$

where $\phi(\mathbf{q}) = \langle\mathbf{q}|\phi\rangle$ and $\psi(\mathbf{q}) = \langle\mathbf{q}|\psi\rangle$ are wave functions (see appendix B).

We have therefore obtained the realization $\mathcal{H} = L^2(\mathbf{R}^k, d^k q)$ for the abstract Hilbert space of states. In physical literature, such a realization is called the *coordinate representation*. The position and momentum operators act in this representation as follows:

$$\hat{\mathbf{q}}\phi(\mathbf{q}) = \mathbf{q}\phi(\mathbf{q}),$$
$$\hat{\mathbf{p}}\phi(\mathbf{q}) = -i\frac{\partial}{\partial\mathbf{q}}\phi(\mathbf{q}).$$

These relations can be written as

$$\langle\mathbf{q}|\hat{\mathbf{q}}|\phi\rangle = \mathbf{q}\langle\mathbf{q}|\phi\rangle,$$
$$\langle\mathbf{q}|\hat{\mathbf{p}}|\phi\rangle = -i\frac{\partial}{\partial\mathbf{q}}\langle\mathbf{q}|\phi\rangle.$$

The passage from the occupation number representation to the coordinate representation is given by

$$\langle\mathbf{n}|\mathbf{q}\rangle = \left(\prod_{i=1}^{k} \bar{H}_{n_i}(q_i)\right)\exp(-\tfrac{1}{2}\mathbf{q}^2),$$

where $\bar{H}_{n_i}$ are normalized Hermite polynomials.

Of course, the functions $\phi_{\mathbf{n}}(\mathbf{q}) = \langle\mathbf{n}|\mathbf{q}\rangle$ form the orthonormal basis of $L^2(\mathbf{R}^k, d^k q)$. Such a basis is clearly the realization of the basis $\{|\mathbf{n}\rangle\}$ in the coordinate representation. In particular, the vacuum vector $|\mathbf{0}\rangle$ is represented by the function

$$\phi_{\mathbf{0}}(\mathbf{q}) = \exp(-\tfrac{1}{2}\mathbf{q}^2),$$

because of the identity $\bar{H}_0(q) \equiv 1$.

The following formula describes the passage from the coherent-state representation to the coordinate representation:

$$\langle\mathbf{z}|\mathbf{q}\rangle = \pi^{-\frac{k}{4}}\exp[\tfrac{1}{2}\mathbf{q}^2 - \tfrac{1}{2}(\mathbf{z}^* - \sqrt{2}\,\mathbf{q})^2 - \tfrac{1}{2}|\mathbf{z}|^2]. \tag{D.5}$$

As an immediate consequence of (D.5), we find that the realization of a normalized coherent state $|\mathbf{z}\rangle$ in the coordinate representation takes the form

$$\phi_{\mathbf{z}}(\mathbf{q}) = \pi^{-\frac{k}{4}}\exp[\tfrac{1}{2}\mathbf{q}^2 - \tfrac{1}{2}(\mathbf{z} - \sqrt{2}\,\mathbf{q})^2 - \tfrac{1}{2}|\mathbf{z}|^2].$$

Furthermore, the formula (D.5) establishes the unitary isomorphism of $L^2(\mathbf{R}^k, d^kq)$ and the Bargmann space $F_2$. In fact, using (D.5), (D.4) and (C.31) we arrive at the following unitary map $\mathfrak{U}$: $\phi \to \tilde{\phi}$ from $L^2(\mathbf{R}^k, d^kq)$ onto $F_2$:

$$(\mathfrak{U}\phi)(\mathbf{z}^*) = \pi^{-\frac{k}{4}} \int d^kq \, \exp[\tfrac{1}{2}\mathbf{q}^2 - \tfrac{1}{2}(\mathbf{z}^* - \sqrt{2}\,\mathbf{q})^2]\phi(\mathbf{q}).$$

On taking the Hermitian conjugate of (D.5) and making use of (C.29) and (C.31) we find that its inverse $\mathfrak{U}^{-1}$: $\tilde{\phi} \to \phi$ is

$$(\mathfrak{U}^{-1}\tilde{\phi})(\mathbf{q}) = \pi^{-\frac{k}{4}} \int_{\mathbf{R}^{2k}} d\mu(\mathbf{z}) \, \exp[\tfrac{1}{2}\mathbf{q}^2 - \tfrac{1}{2}(\mathbf{z} - \sqrt{2}\,\mathbf{q})^2 - |\mathbf{z}|^2]\tilde{\phi}(\mathbf{z}^*).$$

Finally, setting $\mathbf{z} = \frac{1}{\sqrt{2}}(\bar{\mathbf{q}} + i\bar{\mathbf{p}})$ we obtain from (D.5) the following relation:

$$|\langle\bar{\mathbf{q}}, \bar{\mathbf{p}}|\mathbf{q}\rangle|^2 = |\phi_{\bar{\mathbf{q}},\bar{\mathbf{p}}}(\mathbf{q})|^2 = \pi^{-\frac{k}{2}} \exp[-(\mathbf{q} - \bar{\mathbf{q}})^2],$$

where $|\bar{\mathbf{q}}, \bar{\mathbf{p}}\rangle \equiv |\mathbf{z}\rangle$.

Thus the probability density for the coordinates when the system is in the coherent state $|\bar{\mathbf{q}}, \bar{\mathbf{p}}\rangle$, is Gaussian, peaked at $\mathbf{q} = \bar{\mathbf{q}}$. We remark that the function $\phi_{\bar{\mathbf{q}},\bar{\mathbf{p}}}$ is called by engineers a *Gabor wavelet* [57].

We conclude this appendix with a few remarks on the "momentum representation". Consider the eigenvectors $|\mathbf{p}\rangle$, where $\mathbf{p} \in \mathbf{R}^k$, of the momentum operators:

$$\hat{\mathbf{p}}|\mathbf{p}\rangle = \mathbf{p}|\mathbf{p}\rangle.$$

The normalized eigenvectors can be written as

$$|\mathbf{p}\rangle = \pi^{-\frac{k}{4}} \exp(\tfrac{1}{2}\mathbf{p}^2)\exp[\tfrac{1}{2}(\mathbf{a}^\dagger + i\sqrt{2}\,\mathbf{p})^2]|\mathbf{0}\rangle.$$

These vectors form the orthogonal and complete set, namely,

$$\langle\mathbf{p}|\mathbf{p}'\rangle = \delta(\mathbf{p} - \mathbf{p}'),$$
$$\int d^kp \, |\mathbf{p}\rangle\langle\mathbf{p}| = I. \tag{D.6}$$

On taking into account (D.6) we arrive at the realization $\mathcal{H} = L^2(\mathbf{R}^k, d^kp)$ for the abstract Hilbert space of states, with the inner product

$$\langle\phi|\psi\rangle = \int d^kp \, \phi^*(\mathbf{p})\psi(\mathbf{p}).$$

This representation is usually known as the *momentum representation*. The action of the position and momentum operators in this representation has the following form:

$$\hat{\mathbf{q}}\phi(\mathbf{p}) = i\frac{\partial}{\partial\mathbf{p}}\phi(\mathbf{p}),$$
$$\hat{\mathbf{p}}\phi(\mathbf{p}) = \mathbf{p}\phi(\mathbf{p}).$$

The passage from the coordinate representation to the momentum representation is given by

$$\langle \mathbf{q}|\mathbf{p}\rangle = (2\pi)^{-\frac{k}{2}} \exp(i\mathbf{p}\cdot\mathbf{q}). \tag{D.7}$$

On taking the Hermitian conjugate of (D.7) and using (D.4) we obtain the unitary map $\mathfrak{F}$: $\phi \rightarrow \hat{\phi}$, where $\phi(\mathbf{q}) = \langle \mathbf{q}|\phi\rangle$ and $\hat{\phi}(\mathbf{p}) = \langle \mathbf{p}|\phi\rangle$, from $L^2(\mathbf{R}^k, d^kq)$ onto $L^2(\mathbf{R}^k, d^kp)$, which is nothing but the Fourier transform such that

$$(\mathfrak{F}\phi)(\mathbf{p}) = (2\pi)^{-\frac{k}{2}} \int d^kq \exp(-i\mathbf{p}\cdot\mathbf{q})\phi(\mathbf{q}). \tag{D.8}$$

In other words, the coordinate representation and the momentum representation are unitarily equivalent. Clearly, the position and momentum operators are unitarily related to each other by

$$\hat{\mathbf{q}} = \mathfrak{F}\hat{\mathbf{p}}\mathfrak{F}^{-1}, \qquad \hat{\mathbf{p}} = \mathfrak{F}^{-1}\hat{\mathbf{q}}\mathfrak{F}.$$

We finally remark that if the system is in the state described by the wave function $\phi(\mathbf{q})$, then the expression $|\hat{\phi}(\mathbf{p})|^2$ is the probability density for the momenta.

## Appendix E

# FUNCTIONAL DERIVATIVE

This appendix summarizes the basic facts about the functional derivatives. Let $W$ be the space of vector-valued functions on $\mathbf{R}^s$ vanishing rapidly as $|x| \to \infty$, where $x \in \mathbf{R}^s$, and let $I[\mathbf{u}]$ be a functional on $W$. By $I'[\mathbf{v}] \equiv I'(\mathbf{u})[\mathbf{v}]$ we denote the (formal) Gateaux derivative of $I$ at the point $\mathbf{u}$ in the direction $\mathbf{v}$, that is,

$$I'[\mathbf{v}] = \frac{\partial}{\partial \epsilon} I[\mathbf{u} + \epsilon \mathbf{v}]\bigg|_{\epsilon=0} . \tag{E.1}$$

Therefore, $I'$ is an operator that depends on $\mathbf{u}$ and acts on the function $\mathbf{v}$. The *functional derivative* (*gradient*) $\delta I/\delta \mathbf{u}$ of $I$ with respect to $\mathbf{u} \in W$, is defined by

$$I'[\mathbf{v}] = \int d^s x\, \mathbf{v}(x) \cdot \frac{\delta I[\mathbf{u}]}{\delta \mathbf{u}(x)} . \tag{E.2}$$

For example, if $I[u] = \int dx\, u \partial_x^2 u$, then

$$I'[v] = \frac{\partial}{\partial \epsilon} \int dx\, (u + \epsilon v)(\partial_x^2 u + \epsilon \partial_x^2 v)\bigg|_{\epsilon=0} = 2 \int dx\, v \partial_x^2 u,$$

so $\delta I/\delta u = 2\partial_x^2 u$.

REMARK 1. We note that the Gateaux derivative is the infinite-dimensional counterpart of the directional derivative. The corresponding functional derivative is a generalization of a partial derivative to the case of functional spaces. Indeed, let $f$ be a functional on $\mathbf{R}^k$. Using the definition (E.1) we find

$$f'(\mathsf{x})[\mathsf{y}] = \frac{\partial}{\partial \epsilon} f(\mathsf{x} + \epsilon \mathsf{y})\bigg|_{\epsilon=0} = \mathsf{y} \cdot \frac{\partial f}{\partial \mathsf{x}} .$$

It should also be noted that the functional derivatives preserve the fundamental properties of usual derivatives such as linearity or validity of the Leibnitz's rule. ■

REMARK 2. We observe that the functional derivative of a functional $I[u]$ of one variable $u$ can be formally defined as

$$\frac{\delta I[u]}{\delta u(x)} = \frac{\partial}{\partial \epsilon} I[u(x') + \epsilon \delta(x - x')]\bigg|_{\epsilon=0} .$$

This definition follows directly from (E.1) and (E.2). ■

Let us assume that $\mathbf{u}$ depends on a parameter $t$, so that $\mathbf{u} = \mathbf{u}(x, t)$. The definition (E.1) leads to

$$\frac{d}{dt}I[\mathbf{u}] = \int d^s x \, \frac{\delta I}{\delta \mathbf{u}} \cdot \frac{\partial \mathbf{u}}{\partial t}. \tag{E.3}$$

It is clear that (E.3) can be regarded as an alternative definition of the functional derivative. On the other hand, (E.3) seems to be the most convenient tool for calculating functional derivatives (gradients) arising in the theory of nonlinear partial differential equations. For instance, let $I[u] = \int dx \, (-\frac{1}{2}u\partial_x^2 u + u^3)$ be the Hamiltonian for the Korteweg-de Vries equation. On taking the time derivative of $I[u]$ we get

$$\frac{d}{dt}I[u] = \int dx \, \partial_t(-\tfrac{1}{2}u\partial_x^2 u + u^3) = \int dx \, (-\partial_x^2 u + 3u^2)\partial_t u.$$

Hence $\delta I/\delta u = -\partial_x^2 u + 3u^2$. Suppose, in general, that we are given a functional of the form

$$I[\mathbf{u}] = \int d^s x \, \mathfrak{I}(\mathbf{u}),$$

where $\mathfrak{I}(\mathbf{u}) \equiv \mathfrak{I}(\mathbf{u}, D^\alpha \mathbf{u})$, $D^\alpha \mathbf{u} = (D^{\alpha_1}u_1, \ldots, D^{\alpha_k}u_k)$, $\alpha_i$ are multiindices and $D^\beta = \partial^{|\beta|}/\partial x_1^{\beta_1} \cdots \partial x_s^{\beta_s}$. On treating $\mathbf{u}$ as the function that depends on a parameter $t$ and using (E.3) we obtain easily the following relation:

$$\frac{\delta I}{\delta \mathbf{u}} = \frac{\partial \mathfrak{I}}{\partial \mathbf{u}} + \sum_\alpha (-1)^{|\alpha|} D^\alpha \left( \frac{\partial \mathfrak{I}}{\partial D^\alpha \mathbf{u}} \right). \tag{E.4}$$

We now specialize, for brevity, to the case of the functional of one variable $I[u]$. Let us assume that the functional $I[u]$ is analytic, i.e. it can be expanded in the convergent series of the form

$$I[u] = \sum_n \frac{1}{n!} \int d^s x_1 \cdots d^s x_n \, \phi_n(x_1, \ldots, x_n) u(x_1) \cdots u(x_n), \tag{E.5}$$

where the functions $\phi_n(x_1, \ldots, x_n)$ are assumed to be symmetric with respect to the variables $x_1, \ldots, x_n$. Applying the definition (E.1) we find

$$I'[v] = \sum_n \frac{1}{n!} \int d^s x \int d^s x_1 \cdots \int d^s x_n \, v(x) \phi_{n+1}(x, x_1, \ldots, x_n) u(x_1) \cdots u(x_n).$$

Hence, taking into account (E.2) we see that

$$\frac{\delta I[u]}{\delta u(x)} = \sum_n \frac{1}{n!} \int d^s x_1 \cdots \int d^s x_n \, \phi_{n+1}(x, x_1, \ldots, x_n) u(x_1) \cdots u(x_n). \tag{E.6}$$

For example, if $I[u] = \exp\left(\int d^s x\, \phi(x) u(x)\right)$, then

$$\begin{aligned}
\frac{\delta I[u]}{\delta u(x)} &= \frac{\delta}{\delta u(x)} \left( \sum_n \frac{1}{n!} \int d^s x_1 \cdots \int d^s x_n\, \phi(x_1) \cdots \phi(x_n) u(x_1) \cdots u(x_n) \right) \\
&= \sum_n \frac{1}{n!} \int d^s x_1 \cdots \int d^s x_n\, \phi(x) \phi(x_1) \cdots \phi(x_n) u(x_1) \cdots u(x_n) \\
&= \phi(x) \exp\left( \int d^s x\, \phi(x) u(x) \right).
\end{aligned} \tag{E.7}$$

The formula (E.7) is the special case of the general relation

$$\frac{\delta F[u]}{\delta u(x)} = \frac{\delta I[u]}{\delta u(x)} F[u],$$

where $F[u] = \exp(I[u])$.

We note that (E.5) and (E.6) imply

$$I[u] = \sum_n \frac{1}{n!} \int d^s x_1 \cdots d^s x_n \left. \frac{\delta^n I[u]}{\delta u(x_1) \cdots \delta u(x_n)} \right|_{u=0} u(x_1) \cdots u(x_n). \tag{E.8}$$

The series (E.8) is known as a *functional Volterra series*. We also remark that coefficients $\phi_n$ in the expansion (E.5) are, in general, the distributions. For instance, if $I[u] = u(x_0)$, then $I[u] = \int d^s x\, \delta(x - x_0) u(x)$, so

$$\frac{\delta I[u]}{\delta u(x)} = \frac{\delta u(x_0)}{\delta u(x)} = \delta(x - x_0). \tag{E.9}$$

Now let $\mathbf{f}(\mathbf{u}) \equiv \mathbf{f}(\mathbf{u}, D^\alpha \mathbf{u})$ be an analytic vector field on $W$ and let $\mathbf{f}'[\mathbf{g}] \equiv \mathbf{f}'(\mathbf{u})[\mathbf{g}]$ be the Gateaux derivative of $\mathbf{f}$ in the direction $\mathbf{g}$ with respect to $\mathbf{u}$, that is,

$$\mathbf{f}'[\mathbf{g}] = \left. \frac{\partial}{\partial \epsilon} \mathbf{f}(\mathbf{u} + \epsilon \mathbf{g}) \right|_{\epsilon=0}. \tag{E.10}$$

The following identity can be easily obtained from (E.10):

$$\mathbf{f}'[\mathbf{g}] = \mathbf{g} \cdot \frac{\partial}{\partial \mathbf{u}} \mathbf{f} + \sum_\alpha (D^\alpha \mathbf{g}) \cdot \frac{\partial}{\partial D^\alpha \mathbf{u}} \mathbf{f}. \tag{E.11}$$

Motivated by the form of (E.9) we define the functional derivative $\delta \mathbf{f} / \delta \mathbf{u}$ of $\mathbf{f}$ with respect to $\mathbf{u} \in W$ by

$$\mathbf{f}'[\mathbf{g}] = \int d^s x'\, \mathbf{g}(\mathbf{u}(x')) \cdot \frac{\delta}{\delta \mathbf{u}(x')} \mathbf{f}(\mathbf{u}(x)), \tag{E.12}$$

where $\mathbf{g}(\mathbf{u}) \equiv \mathbf{g}(\mathbf{u}, D^\beta \mathbf{u})$.

On using (E.12) and (E.11) we arrive at the following relation:

$$\frac{\delta f_i(\mathbf{u}(x))}{\delta u_j(x')} = \frac{\partial f_i(\mathbf{u}(x))}{\partial u_j(x)}\delta(x-x') + \sum_{\alpha_j} \frac{\partial f_i(\mathbf{u}(x))}{\partial D^{\alpha_j} u_j(x)} D^{\alpha_j}\delta(x-x'), \tag{E.13}$$

where $i, j = 1, \ldots, k$, and $D^{\alpha_j}$ acts on $x$ variable.

REMARK. The reader may notice that the operator

$$X_{\mathbf{g}} = \int d^s x \, \mathbf{g}(\mathbf{u}(x)) \cdot \frac{\delta}{\delta \mathbf{u}(x)} \tag{E.14}$$

given by (E.12) is a natural generalization of the Lie derivative to the case of functional spaces. In fact, let $\mathbf{f}$, $\mathbf{g}$ be vector fields on $\mathbf{R}^k$. Applying the definition (E.10) we get

$$\mathbf{f}'(\mathbf{x})[\mathbf{g}] = \frac{\partial}{\partial \epsilon}\mathbf{f}(\mathbf{x}+\epsilon\mathbf{g})\bigg|_{\epsilon=0} = \mathbf{g}(\mathbf{x})\cdot\frac{\partial}{\partial \mathbf{x}}\mathbf{f}.$$

On the other hand, the operator (E.14) is clearly an infinite-dimensional counterpart of a vector field $\mathbf{g}(\mathbf{x})\cdot\frac{\partial}{\partial \mathbf{x}}$. ■

# BIBLIOGRAPHY


1. B.O. Koopman, *Hamilton systems and transformations in Hilbert space*, Proc. Nat. Acad. Sci. U.S.A. **17** (1931) 315–318.
2. I.P. Kornfeld, Ya. Sinai and S.V. Fomin, *Ergodic Theory* (Nauka, Moscow, 1984) (Russian).
3. T. Carleman, *Application de la théorie des équations intégrales linéaires aux systèmes d'équations différentielles non linéaires*, Acta Mathematica **59** (1932) 63–87.
4. K. Kowalski and W.-H. Steeb, *Nonlinear Dynamical Systems and Carleman Linearization* (World Scientific, Singapore, 1991).
5. S. Fučik and A. Kufner, *Nonlinear Differential Equations* (Elsevier, Amsterdam, 1980).
6. W.S. Wong, *Carleman transformation and Ovcyannikov-Treves operators*, Nonlinear Analysis, TMA **6** (1982) 1295–1308.
7. R. Wackerbauer, W. Eberl, A. Hübler and E. Lüscher, *Analytic representation of stroboscopic maps of ordinary nonlinear differential equations*, H.P.A. **61** (1988) 236–239.
8. R. Wackerbauer, G. Meyer-Kress and A. Hübler, *Algebraic calculation of stroboscopic maps of ordinary, nonlinear differential equations* (1991) preprint.
9. T. Alanson, *A "quantal" Hilbert space formulation for nonlinear dynamical systems in terms of probability amplitudes*, Phys. Lett. A **163** (1992) 41–45.
10. A.C. Newell, *Solitons in Mathematical Physics and Physics* (SIAM, Philadelphia, 1985).
11. T. Nishigori, *Space-time memory functions and solutions of nonlinear evolution equations*, J. Math. Phys. **23** (1982) 2048–2052.
12. K. Kowalski, *Hilbert space description of classical dynamical systems I*, Physica A **145** (1987) 408–424.
13. K. Kowalski, *Hilbert space description of classical dynamical systems II*, Physica A **152** (1988) 98–108.

14. K. Kowalski, *Hilbert space description of nonlinear discrete-time dynamical systems*, Physica A **195** (1993) 137–148.

15. W.-H. Steeb, *Embedding of nonlinear finite dimensional systems in linear infinite dimensional systems and Bose operators*, Hadronic J. **6** (1983) 68–76.

16. L.S. Schulman, *Techniques and Applications of Path Integration* (Wiley, New York, 1981).

17. W. Gröbner and H. Knapp, *Contributions to the Method of Lie Series* (Bibliographisches Institut, Mannheim, 1967).

18. V.I. Arnol'd, *Ordinary Differential Equations* (MIT Press, Cambridge, 1973).

19. C.M. Cheng and P.C.W. Fung, *The evolution operator technique in solving the Schrödinger equation, and its application to disentangling exponential operators and solving the problem of a mass-varying harmonic oscillator*, J. Phys. A: Math. Gen. **21** (1988) 4115–4131.

20. K. Kowalski and W.-H. Steeb, *Symmetries and first integrals for nonlinear dynamical systems: Hilbert space approach I. Ordinary differential equations*, Progr. Theor. Phys. **85** (1991) 713–722.

21. L.C. Biedenharn and J.D. Louck, *Angular Momentum in Quantum Physics. Theory and Application* (Addison-Wesley, Massachusetts, 1981).

22. A.N. Filatov, *Generalized Lie Series and their Applications* (IAN USSR, Tashkent, 1963) (Russian).

23. H. Goldstein, *Classical Mechanics* (Addison-Wesley, Massachusetts, 1980).

24. K. Kowalski, *Hilbert space formulations for nonlinear dynamical systems and Carleman linearization* (1993) preprint.

25. M. Reed and B. Simon, *Methods of Modern Mathematical Physics II: Fourier Analysis, Self-Adjointness* (Academic Press, New York, 1975).

26. V.S. Varadarajan, *Geometry of Quantum Theory*, Vol. II (Van Nostrand Reinhold, New York, 1970) pp. 59–61.

27. M. Reed and B. Simon, *Methods of Modern Mathematical Physics 1. Functional Analysis* (Academic Press, New York, 1972).

28. Yu.G. Pavlenko, *Hamiltonian Methods in Electrodynamics and Quantum Mechanics* (IMU, Moscow, 1985) (Russian).

29. Y. Kano, *A remark on time evolution of coherent states*, Phys. Lett. A **56** (1976) 7–8.

30. A.S. Monin and A.M. Yaglom, *Statistical Fluid Mechanics* (MIT Press, Cambridge, 1971).

31. K. Kowalski and W.-H. Steeb, *Symmetries and first integrals for nonlinear dynamical systems: Hilbert space approach II. Partial differential equations*, Progr. Theor. Phys. **85** (1991) 975–983.

32. W.-H. Steeb, *Linearization procedure and nonlinear systems of differential and difference equations*, in *Nonlinear Phenomena in Chemical Dynamics*, eds. C. Vidal and A. Pacault (Springer, Berlin, 1981) p. 275.

33. F. Schwarz and W.-H. Steeb, *Symmetries and first integrals for dissipative systems*, J. Phys. A: Math. Gen. **17** (1984) L819–L823.

34. M. Kuś, *Integrals of motion for the Lorenz system*, J. Phys. A: Math. Gen. **16** (1983) L689–L691.

35. W.-H. Steeb, A. Kunick and W. Strampp, *The Rikitake two-disc dynamo system and the Painlevé property*, J. Phys. Soc. Jpn. **52** (1983) 2649–2653.

36. A.S. Fokas, *Symmetries and integrability*, Stud. Appl. Math. **77** (1987) 253–299.

37. K. Kowalski, *Linearization transformations for non-linear dynamical systems: Hilbert space approach*, Physica A **180** (1992) 156–170.

38. K. Kowalski, *Linearization of homogeneous Riccati systems*, Phys. Lett. A **109** (1985) 79–80.

39. K. Kowalski, *Bäcklund transformations for nonlinear evolution equations: Hilbert space approach*, Physica A **198** (1993) 493–502.

40. P. Cvitanovic, ed., *Universality in Chaos* (Adam Hilger, Bristol, 1984).

41. J.E. Hirsch, M. Nauenberg and D.J. Scalapino, *Intermittency in the presence of noise: a renormalization group formulation*, Phys. Lett. A **87** (1982) 391–393.

42. K. Kowalski, *A hypothesis concerning "quantal" Hilbert space criterion of chaos in nonlinear dynamical systems* (1994) preprint.

43. P.A.M. Dirac, *The Principles of Quantum Mechanics* (Oxford University Press, London, 1958).

44. S. Klarsfeld and J.A. Oteo, *Recursive generation of higher-order terms in the Magnus expansion*, Phys. Rev. A **39** (1989) 3270–3273.

45. I. Bialynicki-Birula, B. Mielnik and J. Plebański, *Explicit solution of the continuous Baker-Campbell-Hausdorff problem and a new expression for the phase operator*, Ann. Phys. **51** (1969) 187–200.

46. J. von Neumann, *Mathematical Foundations of Quantum Mechanics* (Princeton University Press, Princeton, 1955).

47. J.R. Klauder and E.C. Sudarshan, *Fundamentals of Quantum Optics* (Benjamin, New York, 1968).

48. J.R. Klauder and B.-S. Skagerstam, *Coherent States—Applications in Physics and Mathematical Physics* (World Scientific, Singapore, 1985).

49. A.M. Perelomov, *Generalized Coherent States and their Applications* (Springer, Berlin, 1986).

50. F.A. Berezin and M.A. Shubin, *Schrödinger Equation* (IMU, Moscow, 1983) (Russian).

51. V. Bargmann, *On a Hilbert space of analytic functions and an associated integral transform I*, Comm. Pure Appl. Math. **14** (1961) 187–214.

52. N. Aronszajn, *Theory of reproducing kernels*, Trans. Am. Math. Soc. **68** (1950) 337–404.

53. F.A. Berezin, *Covariant and contravariant symbols of operators*, Izv. AN USSR: Math. Ser. **36** (1972) 1134–1167 (Russian).

54. J. Glimm and A. Jaffe, *Boson quantum field models*, in *Mathematics of Contemporary Physics*, ed. R. Streater (Academic Press, New York, 1972).

55. V. Bargmann, *On a Hilbert space of analytic functions and an associated integral transform II*, Comm. Pure Appl. Math. **20** (1967) 1–101.

56. F.A. Berezin, *The Method of Second Quantization* (Academic Press, New York, 1966).

57. D. Gabor, *Theory of communication*, J. Inst. Elec. Engrs. (London) **93** (1946) 429–457.

## SYMBOL INDEX

| | |
|---|---|
| $\mathbf{N}$ | natural numbers |
| $\mathbf{Z}_+$ | nonnegative integers |
| $\mathbf{R}$ | real numbers |
| $\mathbf{C}$ | complex numbers |
| $\mathbf{n}$ | vector in $\mathbf{Z}_+^k$ |
| $\mathbf{x}$ | vector in $\mathbf{R}^k$ |
| $\mathbf{z}$ | vector in $\mathbf{C}^k$ |
| $\mathbf{e}_i$ | unit vector: $\mathbf{e}_i = (0, \ldots, 0, 1_i, 0, \ldots, 0)$, $i = 1, \ldots, k$ |
| $\mathbf{z} \cdot \mathbf{w}$ | inner product in $\mathbf{C}^k$ |
| $z^*$ | complex conjugate of $z \in \mathbf{C}$ |
| $\mathbf{z}^*$ | vector in $\mathbf{C}^k$ such that $(\mathbf{z}^*)_i = z_i^*$, $i = 1, \ldots, k$ |
| $\lvert\mathbf{z}\rvert$ | norm of $\mathbf{z} \in \mathbf{C}^k$ |
| $\langle \cdot \vert \cdot \rangle$ | inner product on a Hilbert space |
| $\mathcal{H}$ | abstract Hilbert space |
| $L^2(\mathbf{R}^s, d^s x)$ | complex Hilbert space of square integrable functions |
| $L^2_{\mathbf{R}}(\mathbf{R}^s, d^s x)$ | real Hilbert space of square integrable functions |
| $\bigoplus \mathcal{H}_i$ | direct sum of Hilbert spaces |
| $\boldsymbol{\xi}$ | vector-valued function such that $\xi_i \in L^2(\mathbf{R}^s, d^s x)$, $i = 1, \ldots, k$ |
| $\mathbf{u}$ | vector-valued function such that $u_i \in L^2_{\mathbf{R}}(\mathbf{R}^s, d^s x)$, $i = 1, \ldots, k$ |
| $\langle \phi \rvert$ | bra-vector in Dirac notation |
| $\lvert \psi \rangle$ | ket-vector in Dirac notation |
| $\lvert \mathbf{n} \rangle$ | basis vector of the occupation number representation |
| $\lvert \mathbf{z} \rangle$ | coherent state |
| $\lvert \boldsymbol{\xi} \rangle$ | functional coherent state |
| $\lvert \mathbf{q} \rangle$ | common eigenvector of position operators, $\mathbf{q} \in \mathbf{R}^k$ |
| $\lvert \phi \rangle \langle \phi \rvert$ | projection on the state $\lvert \phi \rangle$ |

| | |
|---|---|
| $L^\dagger$ | Hermitian conjugate of the linear operator $L$ |
| Tr | trace |
| $I$ | identity operator |
| $[\cdot,\cdot]$ | commutator |
| $\delta_{ij}$ | Kronecker delta |
| $\delta(x-x')$ | Dirac delta function |
| $a_i$, $a_j^\dagger$ | Bose annihilation and creation operators, $i, j = 1, \ldots, k$ |
| $a_i(x)$, $a_j^\dagger(x')$ | Bose field annihilation and creation operators, $i, j = 1, \ldots, k$ |
| $N_i$ | number operators, $i = 1, \ldots, k$ |
| $\hat{q}_i$, $\hat{p}_j$ | position and momentum operators, $i, j = 1, \ldots, k$ |
| $\frac{\partial}{\partial \mathbf{x}}$ | nabla operator |
| $\partial_t$, $\partial_x$ | shorthand notation for partial derivatives: $\partial_t \equiv \frac{\partial}{\partial t}$, $\partial_x \equiv \frac{\partial}{\partial x}$ |
| $D^\alpha$ | generalized derivative: $D^\alpha = \partial^{\lvert\alpha\rvert}/\partial x_1^{\alpha_1} \cdots \partial x_s^{\alpha_s}$ |
| $D^\alpha \boldsymbol{\xi}$ | vector in $\mathbf{C}^k$ such that $(D^\alpha \boldsymbol{\xi})_i = D^{\alpha_i}\xi_i$, $i = 1, \ldots, k$ |
| $f'[g]$ | Gateaux derivative |
| $\delta/\delta u$ | functional derivative (gradient) |

# SUBJECT INDEX

www.ingramcontent.com/pod-product-compliance
Lightning Source LLC
Chambersburg PA
CBHW050643190326
41458CB00008B/2404
* 9 7 8 9 8 1 0 2 1 7 5 3 2 *